"To my loving family

- my wife, my two sons and their wives, and my five beautiful grandchildren

- Erin, Sean, Leah, Christian and Ryan.

You all bring so much joy and happiness into my life.

This book is dedicated to each and every one of you with all my love."

Table of Contents

Introduction 4
Parts of the Microscope 6
Microscope Activities 8
    Title: "Mastering the Compound Microscope" 8
    Title: "Measuring the Field of View" 10
    Title: "Exploring the Letter 'e' and Slide Movement with a Compound Microscope" 11
    Title: "Smartphone Microscopy: Capturing the Microscopic World" 12
    Title: "Slide into Science: Making a Wet Mount Slide" 13
    Title: "Peeling Back the Layers: Observing Onion Skin Cells" 14
    Title: "Osmosis in Action: Observing Plant Cells" 15
    Title: "Going Bananas: A Microscopic Exploration of Banana Cells" 17
    Title: "Exploring the World of Elodea" 18
    Title: "Light Up Your Life: The Effect of Light on Elodea Chloroplasts" 19
    Title: "Stomata Safari: Exploring the Leaf's Breathing Holes" 20
    Title: "Rooting for Science: Observing Root Hair Cells" 21
    Title: "Pollen Power: A Microscopic Adventure" 23
    Title: "Strawberry Hairs Under the Microscope" 25
    Title: "Orange you curious? A microscopic exploration of citrus cells" 27
    Title: "Going Bananas: A Microscopic Exploration of Banana Cells" 28
    Title: "Cork Cells: A Closer Look" 29
    Title: "Yeast: The Tiny but Mighty Microorganisms" 30
    Title: "Exploring the World of Mushroom Spores" 31
    Title: "Algae Adventure: A Microscopic Exploration" 32
    Title: "Lichen Lookout: A Microscopic Exploration" 34
    Title: "Spice it Up: A Microscopic Exploration of Spices" 36
    Title: "Paper Under the Microscope" 37
    Title: "Hitching a Ride: Observing the Unique Adaptations of Hitchhiker Plants" 38
    Title: "Dust Bunny Detectives" 39
    Title: "Web Weaving: A Microscopic Exploration of Spider Webs" 40
    Title: "Cobweb Wonders: A Microscopic Exploration" 41
    Title: "The Hook and Loop of Velcro: A Microscopic Exploration" 42
    Title: "Ink-credible! Comparing Different Kinds of Print on Paper" 43
    Title: "Sand-tastic! Exploring the Diversity of Sand Under the Microscope" 44
    Title: "Diatom Discovery: Exploring the Microscopic World of Algae" 46
    Title: "Crystal Shapes: A Microscope Activity for Observing Different Types of Crystals" 47

Title: "Bubbling Volcanoes: A Microscopic View of a Chemical Reaction" 49
Title: "Crystal Clear: Comparing the Growth of Different Crystals Under the Microscope" 51
Title: "Fiber Fun: Exploring the Microscopic World of Textile Materials" 53
Title: "Snowflake Science: Exploring the Beauty and Diversity of Snow Crystals 55
Title: "Fishy Scales: Exploring the Structure and Function of Fish Scales" 57
Title: "How old is that fish? A scale-reading activity." 59
Title: "Seeing Red – A Bloody Investigation" 61
Title: "Cheeky Cells: A Microscopic Exploration" 62
Title: "Fingerprint Forensics: A Microscopic Investigation" 63
Title: "Hair Ye Hair Ye : A Microscopic Investigation" 64
Title: "Hair Today, Gone Tomorrow: Examining Hair Damage" 65
Title: "Hair Detective: Solving the Mystery of the Unknown Hair Sample" 66
Title: "Lashing Out: A Close-Up Look at Eyelashes" 67
Title: "Feather Frenzy: Investigating the Structure of Feathers" 68
Title: "Sleepy Bugs: Anesthetizing Insects with Common Materials" 69
Title: "Mosquitoes Up Close: A Compound Microscope Investigation" 70
Title: "Winging It: Investigating Insect Wings with a Compound Microscope" 71
Title: "Flea Frenzy: A Close-Up Look at Fleas" 72
Title: "Fruit Fly Frenzy: Observing Drosophila melanogaster under the Microscope" 73
Title: "Fly Investigation: A Compound Microscope Adventure" 74
Title: "Ants Under the Lens: A Compound Microscope Exploration" 76
Title: "Brine Shrimp and Microscopes: A Closer Look" 78
Title: "Daphnia Discovery: A Microscopic Adventure" 80
Title: "Daphnia's Heartbeat: The Effect of Water Temperature" 82
Title: "Observing Amoeba: A Microscopic Adventure" 84

Introduction

Children thrive on practical learning opportunities. After more than 34 years of instructing students in the science classroom I can attest that some of the most thrilling experiences they ever had were the discoveries they made while using the compound microscope. Their joy was palpable, whether it was simply seeing salt crystals form on a slide for the first time or watching an amoeba devour a nearby organism.

This book includes **57 easy-to-follow microscope activities** that I personally used in the classroom as a science teacher. In addition, **more than 100 science projects** suggestions related to those activities are included.

Each activity is structured to be used in the classroom by teachers and also designed to be used by parents and students independently.

Each activity included in this book includes:

1. An activity title along with a description of the activity's objective(s)
2. A Listing of the materials and the tools required.
3. Step by step instructions
4. Safety precautions
5. Discussion questions: Including open-ended questions related to the activity.
6. Activity Related Science Project Suggestions

If you're teacher looking for engaging STEM activities to incorporate into lesson plans – this book is for you. The book was purposely designed in an 8 x 11-page format to simplify the photocopying of pages for classroom use. If you're a parent wanting to encourage curiosity and a love for learning in your child - this book is for you. And last but not least if you're a student interested in science or searching for a microscope related science project–this book is also for you.

Note: This book was written by George H. Boyd, a former science teacher in the South Huntington School District on Long Island, New York. Mr. Boyd spent the majority of his 34 year teaching career devoted to the science education of gifted children. He presently holds two master's degrees in science - one of which is a master's degree in Science Education. In addition, Mr. Boyd owned and operated the S.M.S. Optical Company established in 1972 for more than 45 years – a company specializing in the sale of microscopes and the servicing of microscopes for schools, industry and the medical profession.

Compound Monocular Microscope

A simple compound microscope is an optical microscope that uses two sets of lenses to magnify small objects that are otherwise too small to see with the naked eye. The two sets of lenses are the objective lenses and the ocular lens (or eyepiece). These lenses work together to produce a magnified image.

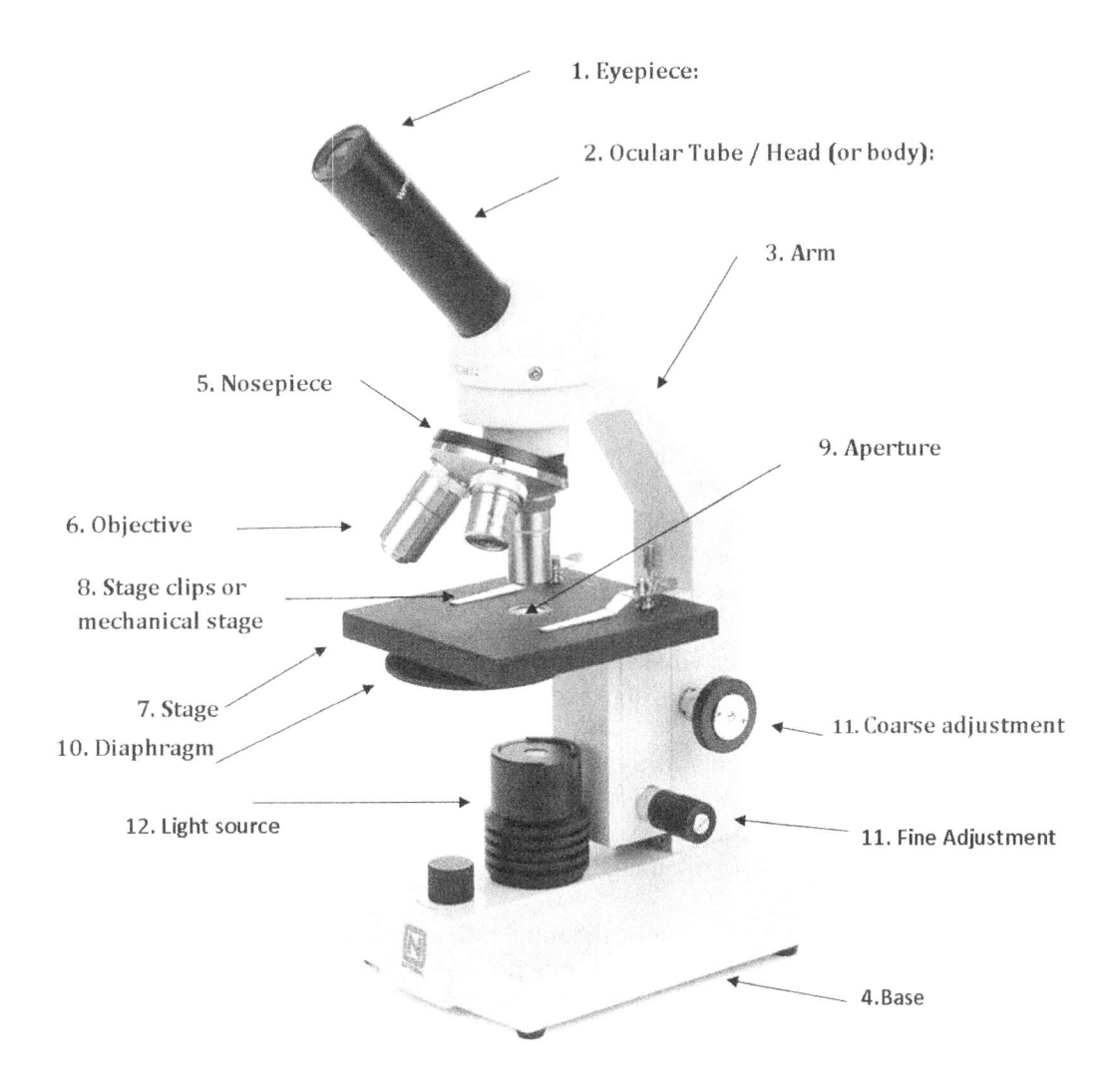

1. Eyepiece:

2. Ocular Tube / Head (or body):

3. Arm

5. Nosepiece

9. Aperture

6. Objective

8. Stage clips or mechanical stage

7. Stage

10. Diaphragm

11. Coarse adjustment

12. Light source

11. Fine Adjustment

4. Base

Parts of the Microscope

- **1. Eyepiece**: A microscope's eyepiece is the component that enlarges the image generated by the instrument's objective lens.. When someone looks through the device, it is typically the lens that is closest to their eye and is also known as an ocular lens. The microscope's overall magnification is calculated by multiplying the eyepiece's own magnification power by the objective lens' own magnification power.

- **2. Ocular Tube / Head (or body)**: The part that holds and connects the eyepiece(s) to the objective lenses. It may be monocular or binocular depending on the number of eyepieces.

- **3. Arm**: The part that supports the head and attaches it to the base of the microscope.

- **4. Base**: The bottom part of the microscope that supports it and houses the light source. The light source could also be a concave mirror that would reflect light up and through the object on the microscope slide.

- **5. Nosepiece**: The part that holds the objective lenses and attaches them to the head. This part rotates to change which objective lens is active.

- **6. Objective lenses**: The lenses closest to the object being viewed. There are usually 3-5 objective lenses on a compound microscope, each with different magnification levels - 4x, 10x, 40x, and 100x are the most common magnifying powers of the objective lenses.

- **7. Stage**: The platform upon which the object or slide is placed. The height of the stage is adjustable on most compound microscopes.

- **8. Stage clips or mechanical stage**: The clips on the stage that hold the slide in place. A mechanical stage allows precise movement of the slide using knobs. The microscope shown has stage clips.

- **9. Aperture**: The circular opening in the stage where the light from the base reaches the object. A disc or iris diaphragm controls the amount of light passing through the aperture.

- **10. Diaphragm**: A disk diaphragm or an iris diaphragm is a device that controls the amount and shape of light passing through the condenser and reaching the slide. A disk diaphragm is a rotating wheel with different-sized holes. The iris diaphragm is a set of blades that can expand or contract to resemble the iris of the eye. The object of both is to obtain the most contrast in the image quality.

- **11. Coarse and fine adjustment controls**: The knobs that adjust the focus of the microscope. The coarse adjustment knob moves the stage up and down quickly, while the fine adjustment knob moves it slowly for precise focusing.

- **12. Light source**: The part that illuminates the object so that it can be viewed through the microscope. It may be a mirror or an electric lamp within the base.

**Total Microscope Magnification
= (Power of the Eyepiece) X (Power of the Objective Lens)**

For Example: If the Eyepiece has a magnifying power of 10X and the Objective Lens has a magnifying power of 40X - together they would produce an image that would make the object appear to be 400 times bigger than its actual size

Required materials and tools:

Listed below are some materials and tools that might not be readily available to complete some of the activities in this book. Possible sources are listed below.

1. Microscope glass slides, well slides, lens paper and coverslips. Available online.

2. Tincture of Iodine: Iodine crystals dissolved in alcohol. Available in drug stores

3 Dissecting Kits containing scalpels, probes, forceps, and tweezers are readily available online.

4. Elodea: This is common pond plant that is also often sold in pet shops for fresh water aquariums.

5. Methylene Blue Solution: methylene blue can be purchased online and in physical stores including science supply companies, pharmacies, pet stores, agricultural supply companies, DIY stores and art media shops.

6. Copper Sulfate: Sold in gardening stores as a fungicide in pet shops and online.

7. Alum - Alum is available in a grocery store as a food additive that is used for pickling and canning. It helps create crisp pickled fruits or vegetables. Alum is usually sold as a fine white powder in the spice aisle or with the pickling ingredients. McCormick is one brand that is readily available.

Microscope Activities

Title: "Mastering the Compound Microscope"

Objectives:

- To learn the parts and functions of a compound microscope
- To understand how to properly use and care for a compound microscope

Materials and Tools:
Compound microscope
Glass slide
Cover slip
Prepared microscope slide (optional)

Instructions:

1. Identify the parts of the compound microscope, including the eyepiece, objective lenses, stage, stage clips, diaphragm, coarse and fine focus knobs, and light source.
2. Learn the function of each part and how they work together to produce a magnified image.
3. Practice using the coarse and fine focus knobs to bring an object into focus.
4. Learn how to properly place a slide on the stage and secure it with the stage clips.
5. Practice changing objective lenses and adjusting the diaphragm to control the amount of light.
6. If using a prepared slide, observe the specimen at different magnifications.

Focusing Procedure:

1. Start with the lowest power objective lens in place.
2. Use the coarse focus knob first to bring the object into focus and then use the fine focus knob to sharpen the image.
3. Once the object is in focus, switch to a higher power objective lens.
4. Use the fine focus knob to bring the object back into focus.

Note: It is important to start with the lowest power objective lens and then switch to higher magnifications. This helps prevent damage to the microscope or specimen.

Safety Precautions:

- Handle the microscope and glass materials with care to avoid breakage.
- Do not touch the lenses with your fingers.

Discussion Questions:

- What are the different parts of a compound microscope and what are their functions?
- How do you properly focus an object using a compound microscope?
- How do you properly place a slide on the stage and secure it with the stage clips?

Notes :

Title: "Measuring the Field of View"

Objectives:

- To learn how to use a compound microscope to measure the size of the field of view.
- To understand how changing the magnification of the microscope affects the size of the field of view.

Materials and Tools:

Compound microscope
Small plastic metric ruler
Lens paper

Instructions:

1. Place the edge of the small plastic metric ruler on the microscope stage with the ruler's edge over the aperture in the middle of the stage.
2. Secure it with the stage clips.
3. Turn on the microscope light and adjust it to a comfortable level.
4. Start with the lowest magnification objective lens and focus on the edge of the ruler.
5. Observe how many millimeters fit within the field of view and record this measurement.
6. Repeat this process for each objective lens, increasing the magnification each time.

Safety Precautions:

- Do not touch the lenses with your fingers; use lens paper to clean them if necessary.

Discussion Questions:

1. How does changing the magnification affect the size of the field of view?
2. Why is it important to know the size of the field of view when using a microscope?

Suggestions For Related Science Projects:

- Investigate how different types of cells (e.g. plant cells, animal cells) appear under different magnifications and measure their sizes.

Title: "Exploring the Letter 'e' and Slide Movement with a Compound Microscope"

Objectives:

- To learn how to use a compound microscope
- To observe the letter "e" on newspaper at different magnifications
- To understand the direction of slide movement as seen through the microscope

Materials and Tools:

- Compound microscope
- Newspaper
- Glass slide
- Cover slip
- Water
- Dropper

Instructions:

1. Cut out a small piece of newspaper with the letter "e" on it.
2. Place the newspaper on the glass slide.
3. Add a drop of water on top of the newspaper using the dropper.
4. Carefully place the cover slip on top of the newspaper, avoiding air bubbles.
5. Place the slide on the microscope stage and secure it with the stage clips.
6. Start observing at the lowest magnification and gradually increase the magnification.
7. While observing, move the slide in different directions and note how the image seen through the microscope moves in the opposite direction.

Safety Precautions:

- Handle the microscope and glass materials with care to avoid breakage.

Discussion Questions:

- How does the appearance of the letter "e" change at different magnifications?
- What other observations can you make about the newspaper at different magnifications?
- How does moving the slide affect what you see through the microscope?

Suggestions For Related Science Projects:

- Observe other materials, such as fabric or plant cells, using a compound microscope.
- Compare and contrast observations made using a compound microscope and a dissecting microscope.

Title: "Smartphone Microscopy: Capturing the Microscopic World"

Objectives:

- Learn how to use a compound microscope
- Learn how to take pictures through a microscope using a smartphone
- Observe and document microscopic specimens

Materials and Tools:

- Compound microscope
- Prepared microscope slides
- Smartphone with camera
- Microscope adapter for smartphone (optional)

Instructions:

1. Set up the compound microscope according to the manufacturer's instructions.
2. Place a prepared microscope slide on the stage and focus the microscope on the specimen.
3. If using a microscope adapter, attach it to the smartphone and align the camera with the microscope eyepiece. If not using an adapter, hold the smartphone camera lens directly over the eyepiece.
4. Adjust the distance of the smartphone from the eyepiece until a clear image appears on the screen of the smartphone. The smartphone camera lens is now at the focal point of the eyepiece.
5. Adjust the focus and lighting as needed to get a clear image on the smartphone screen.
6. Take a picture using the smartphone camera.

Safety Precautions:

- Handle the microscope and slides carefully to avoid damage.
- Follow all manufacturers' instructions for using the microscope.

Discussion Questions:

- What did you observe in your specimen?
- How does using a smartphone to take pictures through a microscope compare to drawing what you see?
- How could this technique be useful in scientific research?

Suggestions For Related Science Projects:

- Create a collection of microscopic images and identify the specimens.
- Compare and contrast different types of cells or tissues using microscopic images.

Title: "Slide into Science: Making a Wet Mount Slide"

A slide is prepared as a wet mound slide when it has a liquid of some kind deposited in the middle of it. This liquid will act as the medium through which the specimen will move or interact with during microscopic inspection.

Objectives:

- To learn how to prepare a wet mount slide
- To understand the importance of using a wet mount slide in microscopy
- To practice proper microscope handling and observation techniques

Materials and Tools:

- Microscope
- Glass slides and cover slips
- Specimen (small piece of tissue paper)
- Dropper
- Water or other mounting medium

Instructions:

1. Place a clean glass slide on a flat surface.
2. Using a dropper, add a drop of water onto the center of the slide.
3. Carefully place your specimen onto the drop of mounting medium.
4. Hold a cover slip at a 45-degree angle and gently lower it onto the specimen, avoiding air bubbles.
5. Observe the slide under the microscope.

Safety Precautions:

- Handle glass slides and cover slips with care as they can break easily.
- Follow proper microscope handling techniques to avoid damaging the instrument.

Discussion Questions:

1. Why is it important to use a wet mount slide in microscopy?
2. What are some common challenges when preparing a wet mount slide?
3. How can you avoid air bubbles when preparing a wet mount slide?

Suggestions For Related Science Projects:

- Observe and compare different types of specimens using wet mount slides
- Investigate the effects of different mounting media on specimen observation
- Explore different staining techniques to enhance specimen contrast

Title: "Peeling Back the Layers: Observing Onion Skin Cells"

An excellent first practice for using a microscope. Under a microscope, the cells are plainly visible, and making a thin section is simple.

Objectives:

- Preparing a wet mount slide
- To observe and identify the basic structure of onion skin cells

Materials and Tools:

- Microscope
- Glass slides and cover slips
- Onion
- Scalpel or sharp knife
- Forceps or tweezers
- Eye Dropper
- Iodine solution (Tincture of Iodine)
- Water

Instructions:

1. Cut a small piece of onion using a scalpel or sharp knife.
2. Carefully peel off a thin transparent layer of the onion skin found between the thicker layers
3. Cut out a small square of onion skin and place it on a clean glass slides
4. Add a drop of water onto the onion skin using a dropper.
5. Add a drop of iodine solution onto the onion skin to stain the cells.
6. Using a tissue absorb some of the water / iodine solution from the slide
7. Carefully place a cover slip over the onion skin, avoiding air bubbles.
8. Observe the slide under the microscope under low power and then under high power.

Note: The iodine solution will stain the cell walls and nuclei of the onion cell

Safety Precautions:

- Be careful when handling sharp objects such as scalpels or knives.
- Wear gloves when handling iodine solution as it may stain skin and clothing.

Discussion Questions:

1. What structures can you identify within the onion skin cells?
2. What is the function of each structure?
3. How does the structure of an onion skin cell compare to other plant cells?

Suggestions For Related Science Projects:

- Observe and compare different types of plant cells
- Investigate the effects of different solutions on onion skin cells

Title: "Osmosis in Action: Observing Plant Cells"

Osmosis is a process by which molecules of a solvent tend to pass through a semipermeable membrane from a less concentrated solution into a more concentrated one, thus equalizing the concentrations on each side of the membrane

Objectives:

- To observe the process of osmosis in plant cells
- To understand the role of the cell membrane in regulating the movement of water

Materials and Tools:

- Microscope
- Microscope slides and cover slips
- Onion
- Iodine solution
- Salt solution
- Distilled water
- Dropper or pipette
- Scalpel or sharp knife
- Tweezers

Note: Glass slides and cover slips are readily available online. Dissecting Kits containing scalpels, probes, forceps, and tweezers are readily available online. Tincture of Iodine : Iodine crystals dissolved in alcohol. Available in drug stores

Instructions:

1. Cut a small piece of onion and peel off a thin layer of the inner epidermis using tweezers.
2. Place the onion epidermis on a microscope slide and add a drop of iodine solution.
3. Carefully place a cover slip over the onion epidermis.
4. Observe the onion cells under the microscope and draw what you see.
5. Add a drop of salt solution to one side of the cover slip and observe any changes in the onion cells.
6. Rinse the slide with distilled water and observe any changes in the onion cells.

Safety Precautions:

- Be careful when handling sharp objects such as scalpels or knives.
- Wear gloves when handling iodine solution.

Discussion Questions:

1. What changes did you observe in the onion cells after adding salt solution?
2. What changes did you observe after rinsing with distilled water?
3. How does this experiment demonstrate the process of osmosis?

Suggestions For Related Science Projects:

- Investigate the effect of different solute concentrations on osmosis in plant cells.
- Compare osmosis in plant cells to osmosis in animal cells.

Notes :

Title: "Going Bananas: A Microscopic Exploration of Banana Cells"

Objectives:

- To learn how to prepare a microscope slide
- To observe the structure of banana skin cells
- To understand the function of different cell structures
- To observe organelles in the cell containing **starch**

Materials and Tools:

- Microscope
- Glass slides and cover slips
- Scalpel or razor blade
- Iodine solution
- Dropper
- Water
- Banana (unripe)

Instructions:

1. Peel a small (tiny) section of skin or pulp from the inside of a banana.
2. Use a scalpel or razor blade to carefully smear it on a slide.
3. Place the slice on a glass slide.
4. Add a drop of water to the slice using a dropper.
5. Add a drop of iodine solution using a dropper
6. Allow the solution to rest for a few moments and the carefully remove some of the liquid from the slide using facial tissue
7. Carefully place a cover slip over the slice.
8. Place the slide on the microscope stage and observe the cells under low power. Starch grains should be visible within the banana cells
9. Adjust the focus and magnification to see the cells more clearly.

Safety Precautions:

- Be careful when using sharp objects such as scalpels or razor blades.
- Use gloves when dealing with an iodine solution. Iodine is capable of staining skin and clothing
- Follow all safety instructions for using a microscope.

Discussion Questions:

1. What structures can you see inside the banana skin cells?
2. What is the function of these structures?
3. How do the cells in the banana skin compare to other plant cells you have observed?

Suggestions For Related Science Projects:

- Compare the structure of banana skin cells to other fruit skin cells.

Title: "Exploring the World of Elodea"

Elodea is a common American water weed commonly found in ponds and readily available in most pet shops

Objectives:

- To observe and identify the cellular structures of the pond plant Elodea
- To understand the function of different cell organelles
- To learn how to prepare and view a wet mount slide

Materials and Tools:

- Microscope
- Glass slides and coverslips
- Dropper or pipette
- Forceps or tweezers
- Scissors
- Elodea plant sample
- Water

Instructions:

1. Cut a small piece of Elodea leaf using scissors.
2. Place the leaf on a clean glass slide.
3. Add a drop of water to the leaf using a dropper or pipette.
4. Carefully place a coverslip over the leaf, avoiding air bubbles.
5. Place the slide on the microscope stage and adjust the focus to view the cells.
6. Observe the cells at different magnifications and identify the cell wall, cell membrane, cytoplasm, chloroplasts, and nucleus. Shine a bright light on the slide – there is a good possibility that you will see the chloroplasts in the Elodea cells circulating within the cytoplasm (the fluid inside the cell)

Safety Precautions:

- Handle the microscope and glass slides with care to avoid breakage.
- Use scissors with caution to avoid injury.

Discussion Questions:

1. What cellular structures can you observe in the Elodea cells?
2. What is the function of each organelle you observed?
3. How does the structure of plant cells differ from animal cells?
4. How does the presence of chloroplasts in Elodea cells relate to its habitat?

Suggestions For Related Science Projects:

- Investigate the effect of different environmental factors on Elodea growth and photosynthesis.
- Compare the cellular structures of different aquatic plants.
- Observe and compare the cellular structures of plant and animal cells.

Title: "Light Up Your Life: The Effect of Light on Elodea Chloroplasts"

Objectives:

- To observe the effect of light on the movement of chloroplasts in Elodea cells
- To understand the role of light in photosynthesis

Materials and Tools:

- Elodea plant
- Microscope
- Glass slides and coverslips
- Dropper
- Water
- Lamp

Instructions:

1. Prepare a wet mount of an Elodea leaf by placing a small piece of the leaf on a glass slide and adding a drop of water. Cover with a coverslip.
2. Observe the Elodea cells under the microscope. Take note of the location and movement of the chloroplasts.
3. Place the slide under a lamp and observe the effect of light on the movement of the chloroplasts.

Safety Precautions:

- Handle the microscope and glass slides with care to avoid injury.

Discussion Questions:

1. What changes did you observe in the movement of the chloroplasts when exposed to light?
2. Why do you think these changes occurred?
3. How does this experiment relate to photosynthesis?

Related Science Projects:

- Investigate the effect of different colors of light on the movement of chloroplasts in Elodea cells.
- Compare the effect of light on chloroplast movement in different types of plants.

Notes:

Title: "Stomata Safari: Exploring the Leaf's Breathing Holes"

Objectives:

- To observe and identify stomata on a plant leaf
- To understand the function of stomata in plants

Materials and Tools:

- Microscope
- Glass slides and cover slips
- Clear nail polish
- Plant leaves (e.g. spinach or ivy)
- Forceps or tweezers
- Toothpicks

Instructions:

1. Paint a small area of the underside of a plant leaf with clear nail polish. Allow to dry completely.
2. Use forceps or tweezers to carefully peel off the dried nail polish, which should have an impression of the leaf's surface.
3. Place the nail polish peel on a glass slide and cover with a cover slip.
4. Observe the slide under the microscope. The stomata should appear as small openings surrounded by guard cells.

Safety Precautions:

- Handle the microscope and glass slides with care to avoid breakage.
- Use caution when handling sharp tools such as forceps or tweezers.

Discussion Questions:

- What is the function of stomata in plants?
- How do environmental factors such as light and humidity affect stomatal opening and closing?
- How might stomatal density vary among different plant species?

Suggestions For Related Science Projects:

- Investigate how different environmental conditions affect stomatal opening and closing.
- Compare stomatal density among different plant species or varieties.
- Explore the relationship between stomatal density and photosynthesis rates.

Title: "Rooting for Science: Observing Root Hair Cells"

Objectives:

- To observe and identify the cells in root hairs
- To understand the function of root hairs in plants
- To practice using a microscope and preparing slides

Materials and Tools:

- Microscope
- Glass slides and cover slips
- Dropper or pipette
- Water
- Plant with visible root hairs (e.g. bean sprout)
- Scalpel or razor blade
- Tweezers

Note: Growing bean sprouts at home is quite simple and can be a fun activity! Here are the steps you can follow to grow your own bean sprouts:

1. Soak the beans overnight: Measure about ½ to ⅔ cup (100 to 133 g) of beans and transfer them to a clean bowl. Cover the beans with fresh water and let them soak for several hours.
2. Drain and rinse the beans: After soaking, drain out the water and rinse the beans until the water runs clear .
3. Place the beans in a container with holes: Transfer the beans to a container with holes and place it in a dark room .
4. Repeat washing and draining every day: Repeat the washing and draining process every day for 5 to 7 days .

At the end of this process, you'll have fresh bean sprouts! It's important to use them quickly as they can spoil in a few days. You can use mung beans or other types of beans for sprouting.

Instructions:

1. Using the tweezers, carefully remove a small section of root with visible root hairs from the plant.
2. Place the root section on a glass slide.
3. Using a dropper or pipette, add a drop of water to the slide.
4. Carefully place a cover slip over the root section, avoiding air bubbles.
5. Place the slide on the microscope stage and observe under low power.
6. Adjust the focus and magnification to observe the cells in the root hairs.

Safety Precautions:

- Handle sharp tools such as scalpels or razor blades with care to avoid injury.
- Follow proper microscope handling procedures to avoid damaging the equipment.

Discussion Questions:

- What do you observe when looking at the cells in the root hairs?
- What is the function of root hairs in plants?
- How do root hairs help plants absorb water and nutrients?

Suggestions For Related Science Projects:

- Investigate how different environmental factors (e.g. soil type, water availability) affect root hair growth and development.
- Compare and contrast root hair cells from different plant species.

NOTES:

Title: "Pollen Power: A Microscopic Adventure"

Objectives:

- To observe and identify pollen grains from different plant species
- To understand the role of pollen in plant reproduction

Materials and Tools:

- Microscope
- Glass slides and cover slips
- Dropper or pipette
- Pollen samples (can be collected from flowers or purchased)
- Tweezers
- Water

Note: There are several methods for collecting pollen. One simple method is to gently tap a flower over a clean glass slide or a piece of paper. The pollen will fall off the anthers (the part of the flower that produces pollen) and onto the slide or paper. You can then use a brush or a pair of tweezers to transfer the pollen to a container for storage or observation.

Another method is to use a small paintbrush or cotton swab to gently brush the anthers and collect the pollen. This method is particularly useful if you want to collect pollen from specific flowers or plant species.

It's important to note that some flowers produce very small amounts of pollen, so you may need to collect from multiple flowers to get enough for observation. Additionally, some plants have sticky pollen that may be more difficult to collect using these methods

Instructions:

1. Set up the microscope according to the manufacturer's instructions.
2. Collect pollen samples using tweezers or by gently tapping a flower over a glass slide.
3. Place a small drop of water on the slide and use a dropper or pipette to transfer the pollen grains onto the water droplet.
4. Carefully place a cover slip over the water droplet, being careful not to trap any air bubbles.
5. Place the slide on the microscope stage and adjust the focus to observe the pollen grains.
6. Compare and contrast the pollen grains from different plant species.

Safety Precautions:

- Be careful when using tweezers or other sharp tools.

Discussion Questions:

- What differences did you observe between the pollen grains of different plant species?
- How do these differences relate to the reproductive strategies of the plants?
- What role do pollinators play in plant reproduction?

Suggestions For Related Science Projects:

- Investigate how different environmental factors (such as temperature or humidity) affect pollen viability.
- Observe and compare pollen grains from different habitats (such as a meadow or a forest).
- Research and create a presentation on the importance of pollinators in ecosystems.

Notes :

Title: "Strawberry Hairs Under the Microscope"

Objectives:

- To learn how to prepare a wet mount slide of a plant specimen.
- To observe the structure and function of strawberry hairs under different magnifications.
- To compare and contrast strawberry hairs with other types of plant hairs.

Materials and Tools:

- Fresh strawberries
- Plain glass microscope slides
- Slide cover slips
- Sharp knife or razor blade
- Water
- Dropper
- Tweezers
- Microscope
- Paper towels
- Petroleum jelly (optional)

Instructions:

1. Carefully cut off a small piece of strawberry skin with a sharp knife or razor blade. Try to include some of the tiny hairs that cover the surface of the strawberry.
2. To make a wet mount of the strawberry skin, put one drop of water in the center of a plain glass slide. The water droplet should be larger than the piece of strawberry skin.
3. Gently place the piece of strawberry skin on top of the water droplet with tweezers. Make sure the hairy side is facing up.
4. Take one cover slip and hold it at an angle to the slide so that one edge of it touches the water droplet on the surface of the slide. Then, being careful not to move the strawberry skin around, lower the cover slip without trapping any air bubbles beneath it. The water should form a seal around the strawberry skin.
5. Use the corner of a paper towel to blot up any excess water at the edges of the cover slip. To keep the slide from drying out, you can make a seal of petroleum jelly around the cover slip with a toothpick.
6. Examine your slide under the microscope. Start with the lowest-power objective and then switch to higher power objectives to see more details. You can also use a magnifying glass to observe your slide before using the microscope.
7. Draw and label what you see on your slide. Try to identify the different parts of the strawberry hairs, such as the base, the shaft and the tip. Note any features that stand out to you, such as the shape, size, color and texture of the hairs.

Safety Precautions:

- Be careful when using sharp knives or razor blades to cut the strawberry skin. Ask an adult for help if needed.
- Do not eat or drink anything in the laboratory area.
- Wash your hands after handling the strawberries and the slides.

Discussion Questions:

- What is the purpose of hairs on plants? How do they help plants survive and adapt to their environment?
- How are strawberry hairs different from other types of plant hairs, such as cactus spines, stinging nettle hairs or cotton fibers?
- How are plant hairs different from animal hairs? What are some similarities and differences in their structure and function?

Suggestions For Related science projects:

- Compare and contrast different types of strawberries under the microscope. How do their hairs vary in shape, size and density?

Notes :

Title: "Orange you curious? A microscopic exploration of citrus cells"

Objectives:

- To observe and identify the different types of cells in an orange
- To understand the structure and function of these cells
- To practice using a microscope and preparing slides

Materials and Tools:

- Orange
- Microscope
- Glass slides and cover slips
- Scalpel or sharp knife
- Dropper or pipette
- Water
- Methylene blue stain (optional)

Instructions:

1. Carefully peel the orange and separate it into segments.
2. Using a scalpel or sharp knife, carefully cut a thin slice from one of the segments.
3. Place the slice on a glass slide and add a drop of water. If using methylene blue stain, add a drop to the slide as well.
4. Carefully place a cover slip over the slice.
5. Observe the slide under the microscope, starting with the lowest magnification and gradually increasing it.

Safety Precautions:

- Be careful when using sharp objects such as scalpels or knives.
- Follow all safety instructions when using methylene blue stain.

Discussion Questions:

- What types of cells did you observe in the orange slice?
- What is the function of these cells?
- How does the structure of these cells relate to their function?

Suggestions For Related Science Projects:

- Compare the cells of different types of citrus fruits (e.g. lemon, lime, grapefruit).
- Investigate how different storage conditions affect the cells of an orange.

Title: "Going Bananas: A Microscopic Exploration of Banana Cells"

Objectives:

- To observe and identify the cells in banana pulp using a microscope
- To understand the basic structure and function of plant cells

Materials and Tools:

- Microscope
- Glass slides and cover slips
- Scalpel or razor blade
- Dropper or pipette
- Water
- Banana

Instructions:

1. Peel a banana and cut a small piece of the pulp using a scalpel or razor blade.
2. Place the banana pulp on a glass slide and add a drop of water.
3. Carefully place a cover slip over the banana pulp.
4. Observe the slide under the microscope and adjust the focus to see the cells clearly.

Safety Precautions:

- Handle the scalpel or razor blade with care to avoid injury.

Discussion Questions:

- What types of cells can you see in the banana pulp?
- How do the cells in banana pulp compare to other plant cells you have observed?
- What is the function of each cell type you observed?

Suggestions For Related Science Projects:

- Compare the cells in different types of fruit.
- Investigate how different treatments (e.g. heating, freezing) affect the cells in banana pulp.

Title: "Cork Cells: A Closer Look"

Cork cells were first observed by an English scientist named Robert Hook over 300 years ago. This was actually the first time that a microscope was used to observe the little box-like structures today called cells.

Objectives:

- To observe the structure of cork cells using a microscope
- To understand the role of cork cells in plants

Materials and Tools:

- Cork
- Razor blade or scalpel
- Microscope
- Microscope slides and cover slips
- Water

Instructions:

1. Using a razor blade or scalpel, carefully cut a thin slice of cork.
2. Place the slice of cork on a microscope slide and add a drop of water.
3. Place a cover slip over the cork slice.
4. Observe the cork slice under the microscope.

Safety Precautions:

- Be careful when handling sharp objects such as razor blades or scalpels.
- Follow all safety guidelines for using a microscope.

Discussion Questions:

- What do you observe about the structure of cork cells?
- How do cork cells differ from other plant cells?
- What is the function of cork cells in plants?

Suggestions For Related Science Projects:

- Compare the structure of cork cells to other plant cells.

Title: "Yeast: The Tiny but Mighty Microorganisms"

Objectives:

- To observe the structure of yeast cells under a microscope
- To learn about the role of yeast in fermentation

Materials and Tools:

- Active dry yeast
- Sugar
- Warm water
- Microscope
- Glass slides and cover slips
- Dropper

Instructions:

1. In a small container, mix a packet of active dry yeast with a teaspoon of sugar and some warm water.
2. Wait for 5-10 minutes until the mixture becomes frothy.
3. Using a dropper, take a small sample of the mixture and place it on a glass slide.
4. Carefully place a cover slip over the sample.
5. Observe the yeast cells under the microscope.

Safety Precautions:

- Handle the microscope and glass slides with care to avoid breakage.

Discussion Questions:

- What structures can you observe on the yeast cells?
- How do these structures help the yeast cells carry out fermentation?
- How do yeast cells compare to other types of cells you have observed?

Suggestions For Related Science Projects:

- Observe and compare different types of yeast under a microscope.
- Investigate how different environmental factors affect the growth and activity of yeast cells.

Title: "Exploring the World of Mushroom Spores"

Objectives:

- To observe and identify mushroom spores under a microscope
- To understand the role of spores in the life cycle of mushrooms

Materials and Tools:

- Microscope
- Glass slides and cover slips
- Mushroom with visible gills or pores (e.g. Agaricus bisporus)

Instructions:

1. Carefully remove the stem from the mushroom cap and place the cap gills/pores side down on a glass slide.
2. Place a cup over the slide to reduce draft and allow to sit overnight
3. Remove the mushroom cap and place a cover slip over the spores collected on the slide,
4. observe the spore print left on the slide under the microscope.

Safety Precautions:

- Always handle the microscope and glass slides with care to avoid injury.
- Wear gloves when handling mushrooms to avoid skin irritation.

Discussion Questions:

1. What do you observe when looking at the spores under the microscope?
2. How do the spores help mushrooms reproduce?
3. How do different types of mushrooms release their spores?

Suggestions For Related Science Projects:

- Investigate how environmental factors such as temperature and humidity affect spore release in mushrooms.
- Compare the spore prints of different types of mushrooms.

Title: "Algae Adventure: A Microscopic Exploration"

Objectives:

- To observe and identify different types of algae using a microscope
- To understand the basic structure and characteristics of algae
- To learn how to prepare wet mount slides for microscopic observation

Materials and Tools:

- Microscope
- Glass microscope slides
- Cover slips
- Dropper or pipette
- Water (preferably pond water)
- Algae sample (can be collected from a pond or aquarium)
- Tweezers
- Paper towels

Step-by-step instructions:

1. Collect a sample of algae from a pond or aquarium using a dropper or pipette. If you don't have access to a natural source of algae, you can purchase a sample from a science supply store. One supply online is "Algae Research Supply.. Here you can buy pure sample of algae and also instructions and nutrients to grow your own.
2. Place a drop of water on the center of a clean glass microscope slide.
3. Using tweezers carefully transfer a small amount of the algae sample onto the drop of water on the slide.
4. Place a cover slip on top of the algae sample, being careful to avoid trapping air bubbles.
5. Blot away any excess water around the edges of the cover slip using a paper towel.
6. Place the slide on the microscope stage and adjust the focus to observe the algae at different magnifications.

Safety precautions:

- Be careful when handling glass microscope slides and cover slips as they can break and cause injury.
- Wash your hands thoroughly after handling the algae sample.

Discussion questions:

- What different types of algae did you observe?
- What are some characteristics that all algae share?
- How do algae differ from other types of plants?

Suggestions For Related science projects:

- Investigate how different environmental factors (such as light, temperature, and nutrient availability) affect the growth of algae.
- Compare and contrast the structure and characteristics of different types of algae (such as green, red, and brown algae).
- Explore the role of algae in aquatic ecosystems and food chains.

Notes :

Title: "Lichen Lookout: A Microscopic Exploration"

Lichens are not plant or animal but a unique combination of an algae and a fungus. or a cyanobacteria and a fungus. They are found worldwide on various types of substrata, Lichens have many uses as food, medicines and dyes.

Objectives:

- To observe and identify different types of lichen using a microscope
- To understand the basic structure and characteristics of lichen
- To learn how to prepare dry mount slides for microscopic observation

Materials and Tools:

- Microscope
- Glass microscope slides
- Cover slips
- Lichen sample (can be collected from trees, rocks, or soil)
- Tweezers
- Paintbrush or soft-bristled brush

Instructions:

1. Collect a sample of lichen from a tree, rock, or soil using tweezers. Be sure to collect only a small amount and avoid damaging the lichen or its substrate.
2. Place the lichen sample on a clean glass microscope slide.
3. Use a paintbrush or soft-bristled brush to gently remove any debris or dirt from the lichen sample.
4. Place a cover slip on top of the lichen sample.
5. Place the slide on the microscope stage and adjust the focus to observe the lichen at different magnifications.

Safety precautions:

- Be careful when handling glass microscope slides and cover slips as they can break and cause injury.
- Wash your hands thoroughly after handling the lichen sample.

Discussion questions:

- What different types of lichen did you observe?
- What are some characteristics that all lichens share?
- How do lichens differ from other types of plants?

Suggestions For Related science projects:

- Investigate how different environmental factors (such as light, temperature, and air quality) affect the growth of lichens.
- Compare and contrast the structure and characteristics of different types of lichens (such as crustose, foliose, and fruticose lichens).
- Explore the role of lichens in terrestrial ecosystems and their importance as bio indicators.

Notes :

Title: "Spice it Up: A Microscopic Exploration of Spices"

Objectives:

- To observe and compare the microscopic structures of different spices
- To learn about the properties and uses of common spices

Materials and Tools:

- Microscope
- Glass slides and cover slips
- Dropper
- Water
- Spices (e.g. black pepper, cinnamon, cumin, nutmeg)

Instructions:

1. Choose a spice to observe and place a small amount on a glass slide.
2. Add a drop of water to the spice using a dropper.
3. Carefully place a cover slip over the spice and water.
4. Place the slide on the microscope stage and observe under low power.
5. Repeat with other spices and compare their microscopic structures.

Safety Precautions:

- Handle the microscope and glass slides with care to avoid damage or injury.
- Wash hands after handling spices to avoid irritation.

Discussion Questions:

- How do the microscopic structures of different spices compare?
- What properties or uses do these spices have?

Suggestions For Related Science Projects:

- Investigate the antimicrobial properties of different spices.
- Explore the history and cultural significance of spices.

Title: "Paper Under the Microscope"

-
- A Close-Up Look at Different Types of Paper
- Paper is actually made from a variety of different materials – wood, cotton, rice straw, wheat straw, old T-shirts, fruit skin, Poo form herbivores, seaweed, leather and synthetic fibers.

Objectives:

- To observe and compare the microscopic structure of different types of paper
- To understand how the microscopic structure of paper affects its properties and uses

Materials and Tools:

- Microscope
- Slides and cover slips
- Dropper or pipette
- Water
- Samples of different types of paper (e.g. toilet paper, printer paper, construction paper, newspaper)

Instructions:

1. Prepare a wet mount slide for each type of paper by placing a small piece of torn paper on a slide and adding a drop of water. Cover with a cover slip.
2. Observe each slide under the microscope at low, medium, and high magnification.
3. Record your observations and compare the microscopic structure of each type of paper.

Safety Precautions:

- Handle the microscope and glass slides with care to avoid breakage.
- Follow all safety guidelines for using a microscope.

Discussion Questions:

- How does the microscopic structure of each type of paper differ?
- How do these differences affect the properties and uses of each type of paper?
- Can you think of any other factors that might affect the properties and uses of paper?

Suggestions For Related Science Projects:

- Investigate how different types of paper react to water or other liquids.
- Explore how different types of paper are made and what materials are used in their production

Title: "Hitching a Ride: Observing the Unique Adaptations of Hitchhiker Plants"

Objectives:

- To observe and identify the unique adaptations of hitchhiker plants that allow their seeds to stick to clothing and fur.
- To understand the role of these adaptations in seed dispersal.

Materials and Tools:

- Microscope
- Slides and cover slips
- Forceps
- Dropper
- Water
- Seeds from various hitchhiker plants (e.g. "Stick-tight" Harpagonella, "Beggerticks", Krameria, Puncturevine, Jumping cholla, Hedge-parsley, Calico aster, Common burdock)

Instructions:

1. Collect seeds from various hitchhiker plants.
2. Place a seed on a microscope slide using forceps.
3. Add a drop of water to the seed using a dropper.
4. Place a cover slip over the seed.
5. Observe the seed under the microscope at different magnifications.
6. Identify and record any unique adaptations that may allow the seed to stick to clothing or fur.
7. Repeat steps 2-6 for each type of seed.

Safety Precautions:

- Handle the microscope and other tools with care.
- Be careful when handling seeds as some may have sharp edges or barbs.

Discussion Questions:

1. What unique adaptations did you observe in the seeds of hitchhiker plants?
2. How do these adaptations help the seeds stick to clothing or fur?
3. How does this aid in seed dispersal?
4. Can you think of any other ways that plants disperse their seeds?

Suggestions For Related Science Projects:

1. Investigate the effectiveness of different methods for removing hitchhiker plant seeds from clothing or fur.
2. Conduct a field study to observe how animals aid in seed dispersal for hitchhiker plants.
3. Research and compare the seed dispersal mechanisms of different plant species.

Title: "Dust Bunny Detectives"

"Dust Bunnies" are small clumps of dust that form under furniture and in corners that are not cleaned regularly. They are made of spider webs, dead skin, dust, and sometimes debris held together by static electricity.

Objectives:

- To collect and observe dust bunnies under a microscope
- To identify the components of dust bunnies
- To understand the formation and accumulation of dust bunnies

Materials and Tools:

- Microscope
- Glass slides and cover slips
- Tweezers
- Gloves
- Dust mask
- Dust bunnies (collected from various locations)

Instructions:

1. Put on gloves and a dust mask to protect yourself from inhaling any particles.
2. Use tweezers to carefully pick up a dust bunny and place it on a glass slide.
3. Place a cover slip over the dust bunny.
4. Place the slide on the microscope stage and observe the dust bunny under low magnification.
5. Increase the magnification to observe the individual components of the dust bunny.
6. Repeat with dust bunnies collected from different locations.

Safety Precautions:

- Wear gloves and a dust mask to avoid inhaling any particles.
- Handle glass slides and cover slips with care to avoid injury.

Discussion Questions:

1. What components did you observe in the dust bunnies?
2. Did you notice any differences between dust bunnies collected from different locations?
3. How do you think dust bunnies form and accumulate?

Suggestions For Related Science Projects:

- Investigate the effectiveness of different cleaning methods in reducing dust bunny accumulation.
- Compare the composition of dust bunnies collected from homes with pets versus homes without pets.

Title: "Web Weaving: A Microscopic Exploration of Spider Webs"

Objectives:

- To collect and prepare a spider web sample for microscopic observation
- To observe and analyze the structure of a spider web under a microscope
- To learn about the properties and uses of spider silk

Materials and Tools:

- Compound microscope
- Microscope slides
- Cover slips
- Clear nail polish
- Tweezers
- Gloves
- Notebook and pen/pencil

Step-by-step Instructions:

1. Find a complete, dry spider web. Take note of the spider on the web (if it is still there).
2. Put on gloves to protect your hands.
3. Paint the center of a microscope slide with clear nail polish.
4. Holding the slide firmly with tweezers, gently touch the slide onto an interesting part of the spider web to collect some of the web.
5. Remove any excess spider web using tweezers.
6. Without delay, cover the slide with a cover slip, pressing it down gently.
7. Place the slide on the microscope stage and observe the spider web under different magnifications.
8. Record your observations in your notebook.

Safety Precautions:

- Always wear gloves when handling spider webs to protect your hands.
- Be careful not to harm or disturb the spider when collecting the web.

Discussion Questions:

1. What did you observe about the structure of the spider web under the microscope?
2. How does the structure of the web relate to its function?
3. What are some properties and uses of spider silk?
4. How do different types of spiders use their webs?

Suggestions For Related Science Projects:

1. Research and compare the properties of spider silk from different species of spiders.
2. Investigate how environmental factors (such as temperature and humidity) affect the properties of spider silk.
3. Design an experiment to test the strength and elasticity of spider silk.

Title: "Cobweb Wonders: A Microscopic Exploration"

Cobwebs are actually sticky spider webs that have been abandoned by Cobb spiders Cob spiders are mainly from the species Therididae. Over time the spider webs weaken, accumulate dust and the spider leaves them..

Objectives:

- To observe and identify the structure and composition of cobwebs using a microscope.
- To understand the role of cobwebs in the ecosystem.

Materials and Tools:

- Microscope
- Glass slides and cover slips
- Tweezers
- Cobweb samples
- Dropper
- Water

Instructions:

1. Collect cobweb samples using tweezers and place them on a glass slide.
2. Add a drop of water to the sample using a dropper.
3. Carefully place a cover slip over the sample.
4. Observe the sample under the microscope at different magnifications.

Safety Precautions:

- Handle the microscope and glass slides with care to avoid breakage.
- Wear gloves while collecting cobweb samples.

Discussion Questions:

- What did you observe about the structure and composition of the cobweb?
- How do cobwebs benefit the ecosystem?

Suggestions For Related Science Projects:

- Investigate the properties of spider silk.
- Observe and compare different types of spider webs.

Title: "The Hook and Loop of Velcro: A Microscopic Exploration"

Velcro is a fastener that consists of two strips of fabric, one with tiny hooks and one with tiny loops. It was invented by a Swiss engineer named George de Mestral in the 1950,s. The plant that gave rise to Velcro was the burdock plant. Its seeds have tiny hooks that attach to animal fur.

Objectives:

- To observe and understand the structure of Velcro using a microscope
- To learn about the properties of Velcro and how it works

Materials and Tools:

- Velcro strip
- Microscope
- Glass slides
- Cover slips
- Tweezers
- Scissors

Instructions:

1. Cut a small piece of Velcro using scissors.
2. Separate the hook and loop sides of the Velcro.
3. Place each side on a glass slide and cover with a cover slip.
4. Observe the hook and loop sides under the microscope at different magnifications.
5. Draw or take pictures of what you see.

Safety Precautions:

- Be careful when using scissors to cut the Velcro.
- Handle the glass slides and cover slips with care to avoid breakage.

Discussion Questions:

1. How does the structure of Velcro allow it to stick together?
2. What are some other materials that have similar properties to Velcro?
3. How does the strength of Velcro compare to other fasteners such as zippers or buttons?

Suggestions For Related Science Projects:

- Experiment with different materials to see which ones stick to Velcro.
- Test the strength of Velcro by attaching weights to it and seeing how much it can hold.
- Compare the structure of natural materials such as burrs to that of Velcro.

Title: "Ink-credible! Comparing Different Kinds of Print on Paper"

Objectives:

- To observe and compare the characteristics of different kinds of print on paper using a compound microscope.
- To understand how different printing techniques affect the appearance of text and images on paper.

Materials and Tools:

- Compound microscope
- Slides and cover slips
- Water
- Dropper
- Various samples of printed paper (e.g. newspaper, magazine, book, inkjet printout, laser printout)

Instructions:

1. Choose a small section of text or an image from one of the paper samples and carefully cut it out.
2. Place the cut-out section onto a microscope slide and add a drop of water on top.
3. Carefully place a cover slip over the sample, making sure to avoid trapping air bubbles.
4. Place the slide onto the microscope stage and observe the sample under low magnification. Adjust the focus and lighting as needed.
5. Repeat steps 1-4 for each paper sample, making sure to clean the slide and cover slip between each use.
6. Compare the appearance of the different samples under the microscope.

Safety Precautions:

- Handle the microscope and glass slides/cover slips with care to avoid breakage.
- Use caution when handling sharp objects such as scissors or blades.

Discussion Questions:

1. How do the different kinds of print on paper differ in appearance under the microscope?
2. What factors might affect the appearance of print on paper (e.g. printing technique, paper quality)?
3. How might this activity be useful in real-world applications (e.g. forensic science)?

Suggestions For Related Science Projects:

- Investigate how different types of ink (e.g. water-based, oil-based) affect the appearance of print on paper under a microscope.
- Compare the appearance of handwritten text versus printed text under a microscope.

Title: "Sand-tastic! Exploring the Diversity of Sand Under the Microscope"

Objectives:

- To observe and compare the physical characteristics of sand samples from different sources
- To learn about the origin and composition of sand grains
- To practice using a compound microscope and preparing slides

Materials and Tools:

- Compound microscope
- Glass slides and cover slips
- Dropper
- Water
- Toothpick
- Sand samples from different locations (e.g., beach, desert, river, playground, etc.)
- Magnifying glass
- Paper and pencil
- Optional: camera or smartphone to take pictures of the sand grains

Instructions:

1. Collect sand samples from different locations. Label each sample with its source and put it in a small container.
2. Examine each sand sample with a magnifying glass. Note the color, shape, size, and texture of the sand grains. Record your observations in a table or a chart.
3. Choose one sand sample to observe under the microscope. Place a small amount of sand on a glass slide and spread it evenly with a toothpick. Add a drop of water to moisten the sand and cover it with a cover slip. Be careful not to trap air bubbles under the cover slip.
4. Place the slide on the microscope stage and adjust the light source. Start with the lowest magnification objective lens (4x) and focus on the sand grains. Observe the sand grains and note their features, such as color, shape, size, texture, and transparency. You may see some sand grains that are transparent, opaque, shiny, dull, smooth, rough, round, angular, or irregular. You may also see some sand grains that are composed of different minerals or materials, such as quartz, feldspar, mica, shell fragments, coral pieces, or volcanic rock.
5. Switch to a higher magnification objective lens (10x or 40x) and observe the sand grains in more detail. You may see some features that were not visible at lower magnification, such as cracks, holes, patterns, or layers in the sand grains. You may also see some microscopic organisms or debris in the sand sample, such as diatoms, algae, pollen, or dust.
6. Draw or take pictures of the sand grains that you observe under the microscope. Label the features that you notice and indicate the magnification level.
7. Repeat steps 3 to 6 for each sand sample that you want to observe under the microscope. Compare and contrast the sand samples from different sources and explain how their characteristics reflect their origin and composition.

Safety Precautions:

- Handle the microscope with care and follow the instructions for its use and maintenance.
- Do not touch the microscope lenses with your fingers or any other objects.
- Do not use force or excessive pressure when adjusting the microscope knobs or moving the slide on the stage.
- Do not look at the light source directly or point it at someone's eyes.
- Wash your hands after handling the sand samples and dispose of them properly.

Discussion Questions:

- What are some factors that affect the color, shape, size, and texture of sand grains?
- How can you tell if a sand grain is composed of one or more minerals or materials?
- How can you identify some common minerals or materials in sand grains?
- What are some sources of variation or error in your observations?
- How can you improve your technique or accuracy in observing sand grains under the microscope?

Suggestions For Related science projects:

- Collect sand samples from different beaches around the world and compare their characteristics under the microscope.
- Investigate how different environmental factors (e.g., wind, water, temperature) affect the erosion and deposition of sand grains.
- Experiment with different methods of separating or sorting sand grains based on their physical properties (e.g., density, magnetism, solubility).
- Explore how different types of sand affect the growth of plants or animals.

Notes :

Title: "Diatom Discovery: Exploring the Microscopic World of Algae"

Diatomaceous earth is produced from a naturally occurring sedimentary rock that is made from fossilized tiny, aquatic organisms called diatoms. This rock is crumbled into powder The powder is actually diatom skeletons made of silica.

Objectives:

- Learn about diatoms and their role in aquatic ecosystems
- Observe diatoms using a compound microscope
- Identify different species of diatoms
- Understand the importance of diatoms as bioindicators of water quality

Materials and Tools:

- Compound microscope
- Glass slides and coverslips
- Dropper or pipette
- Diatomaceous earth (Often used as a filtering material in pool systems)
- Glass of water

Instructions:

1. Obtain a sample of powdered diatomaceous earth.
2. Mix the earth is water creating a dense cloudy liquid.
3. Place a drop of the water sample onto a glass slide and cover with a coverslip.
4. Place the slide onto the stage of the compound microscope and observe under low magnification.
5. Adjust the focus and magnification to observe the diatoms in greater detail.

Safety Precautions:

- Handle the microscope and glass slides with care to avoid damage or injury.
- Wash hands thoroughly after handling the water sample.

Discussion Questions:

1. What are some characteristics that can be used to identify different species of diatoms?
2. Why are diatoms important for aquatic ecosystems?
3. How can diatoms be used as bio indicators of water quality?

Suggestions For Related Science Projects:

- Investigate the effects of different environmental factors (such as temperature, pH, and nutrient levels) on diatom populations.
- Compare diatom populations in different bodies of water to assess their water quality.
- Conduct a long-term study to monitor changes in diatom populations over time.

Title: "Crystal Shapes: A Microscope Activity for Observing Different Types of Crystals"

Objectives:

- To learn how to prepare and observe microscope slides of different types of crystals.
- To compare and contrast the shapes, sizes, colors, and structures of various crystals.
- To understand how the chemical composition and crystal structure of substances affect their physical properties.

Materials and Tools:

- Microscope
- Glass slides and cover slips
- Pipettes or droppers
- Distilled water
- Lens paper
- Toothpicks
- Petroleum jelly
- Beakers or cups
- Stirring rods or spoons
- Hot plate or microwave
- Sugar
- Salt
- Alum
- Copper sulfate

Instructions:

1. Prepare four solutions of sugar, salt, alum, and copper sulfate by dissolving about a teaspoon of each substance in 100 mL of distilled water in separate beakers or cups. Stir well until no more solid can dissolve. You may need to heat the water slightly to help dissolve the substances. Label each beaker or cup with the name of the substance.
2. Take a glass slide and put one drop of one of the solutions in the center. Gently place a cover slip over the drop, making sure there are no air bubbles trapped under it. You can use a toothpick to slide the cover slip over the drop. If there is any excess liquid at the edges of the cover slip, you can blot it with a paper towel. To prevent evaporation, you can seal the edges of the cover slip with petroleum jelly using a toothpick.
3. Repeat step 2 for each of the other solutions, using a new slide and cover slip for each one. Label each slide with the name of the substance.
4. Place one of the slides on the microscope stage and secure it with the clips. Turn on the microscope light and adjust the condenser and diaphragm to get maximum illumination. Start with the lowest power objective (4X) and focus on the drop using the coarse and fine adjustment knobs. Observe the crystals that have formed in the drop and note their shape, size, color, and structure. You can draw or take a picture of what you see.
5. Switch to a higher power objective (10X or 40X) and refocus on the crystals using only the fine adjustment knob. Observe any details that you could not see with the lower power objective. Note any differences in crystal habit, growth zones, nucleation, or deformation textures.
6. Repeat steps 4 and 5 for each of the other slides, using a different objective for each one. Compare and contrast your observations for each type of crystal.
7. When you are done observing, turn off the microscope light and lower the stage. Remove the slide from the stage and clean it with lens paper moistened with alcohol. Wipe off any oil or petroleum jelly from the slide and

cover slip. Dispose of the slide and cover slip in a designated container or trash bin. Do not reuse them for another activity.
8. Clean your microscope lenses with lens paper moistened with alcohol. Wipe off any oil or petroleum jelly from the lenses. Return your microscope to its original condition and place.

Safety Precautions:

- Wear safety goggles, gloves, and aprons when handling chemicals and hot liquids.
- Do not ingest or inhale any of the substances used in this activity.
- Wash your hands thoroughly after completing this activity.
- Be careful when using sharp objects such as knives, razors, or toothpicks.
- Be careful when using electrical appliances such as hot plates or microwaves.
- Follow your instructor's directions for disposing of waste materials.

Discussion Questions:

- What are crystals and how do they form?
- What factors affect the shape

Suggestions For Related science projects:

- Collect a variety of crystals from your home environment and describe their shapes
- Investigate how to grow crystals Example . Rock Candy

Notes :

Title: "Bubbling Volcanoes: A Microscopic View of a Chemical Reaction"

Objectives:

- To observe the reaction of baking soda and vinegar under a microscope
- To understand the chemical equation and the types of reactions involved
- To explore the effects of changing the amounts of reactants and temperature on the reaction rate

Materials and Tools:

- Baking soda (sodium bicarbonate)
- Vinegar (acetic acid solution)
- Microscope
- Microscope slides
- Cover slips
- Droppers
- Small cups or beakers
- Thermometer
- Stopwatch
- Paper towels

Instructions:

1. Place a small amount of baking soda on a microscope slide and cover it with a cover slip.
2. Place the slide on the microscope stage and adjust the focus to see the baking soda crystals clearly.
3. Use a dropper to add a drop of vinegar to the edge of the cover slip and observe what happens under the microscope. Record your observations and start the stopwatch.
4. Continue to observe the reaction until it stops or slows down significantly. Record the time elapsed and any changes you notice in the appearance of the baking soda and vinegar.
5. Repeat steps 1 to 4 with different amounts of baking soda and vinegar (e.g., more baking soda, less vinegar, or vice versa) and compare the results. How does changing the amounts of reactants affect the reaction rate and the amount of products formed?
6. Repeat steps 1 to 4 with vinegar at different temperatures (e.g., cold, room temperature, or warm) and compare the results. How does changing the temperature affect the reaction rate and the amount of products formed?

Safety Precautions:

- Wear safety goggles and gloves when handling vinegar and baking soda.
- Do not touch or inhale the gas produced by the reaction, as it may irritate your eyes, nose, or throat.
- Do not ingest or taste any of the materials used in this activity.
- Clean up any spills or messes with paper towels and dispose of them properly.

Discussion Questions:

- What did you see under the microscope when you added vinegar to baking soda? What were those bubbles made of?
- What type of reaction is happening between baking soda and vinegar? What are the reactants and products in this reaction? Write down the balanced chemical equation for this reaction.
- How can you tell that a chemical reaction has occurred? What are some signs or evidence of a chemical change?
- How did changing the amounts of baking soda and vinegar affect the reaction rate and the amount of products formed? Explain your observations using the concept of limiting reactant.
- How did changing the temperature of vinegar affect the reaction rate and the amount of products formed? Explain your observations using the concept of kinetic energy and collision theory.

Suggestions For Related science projects:

- Make a model volcano using clay or paper mache and use baking soda and vinegar to simulate an eruption. Measure how high the "lava" shoots up and how long it lasts. Experiment with different amounts and concentrations of baking soda and vinegar to see how they affect the eruption.
- Make "hot ice" by boiling off or evaporating all the water from a baking soda and vinegar solution. The resulting solid is sodium acetate, which can be used to make instant ice sculptures by touching it with another piece of sodium acetate or a metal object.

Notes :

Title: "Crystal Clear: Comparing the Growth of Different Crystals Under the Microscope"

Objectives:

- To learn how to prepare and observe different types of crystals under the microscope
- To compare and contrast the shape, size, color and structure of salt, sugar and copper sulfate crystals
- To identify the factors that affect the growth and quality of crystals

Materials and Tools:

- Three clear glass jars with lids
- Distilled water
- Table salt
- Sugar
- Copper sulfate
- Measuring spoons
- Stirring rods
- Microscope slides and coverslips
- Droppers or straws

Instructions:

1. Label the jars as salt, sugar and copper sulfate. Fill each jar with 200 ml of distilled water.
2. Add four tablespoons of table salt to the salt jar, four tablespoons of sugar to the sugar jar, and two tablespoons of copper sulfate to the copper sulfate jar. Stir each solution well until most of the solid dissolves. You may see some undissolved solid at the bottom of the jars, which is normal.
3. Place the jars in a sunny spot and leave them undisturbed for several days until most of the water evaporates. You should see some crystals forming on the sides and bottom of the jars.
4. Carefully remove the lids from the jars and use a dropper or a straw to transfer some of the remaining liquid to a microscope slide. Cover it with a coverslip and place it on the microscope stage.
5. Observe the slide under low power and then high power objective lenses.. Draw and label what you see Note the color, shape, size and arrangement of the crystals. Repeat this step for each type of crystal.
6. Use a stirring rod or a tweezers to gently scrape off some of the crystals from the sides or bottom of the jars. Place them on a microscope slide and cover them with a coverslip. Observe them under low power and then high power objective lenses. Draw and label what you see . Note the color, shape, size and arrangement of the crystals. Repeat this step for each type of crystal.

Safety Precautions:

- Wear gloves and goggles when handling copper sulfate, as it can irritate your skin and eyes.
- Do not ingest any of the solutions or crystals, as they can be harmful if swallowed.
- Wash your hands thoroughly after handling any of the materials.
- Dispose of the solutions and crystals

Discussion Questions:

- How do salt, sugar and copper sulfate crystals differ in their appearance under the microscope?
- What are some similarities among salt, sugar and copper sulfate crystals?
- How do you think the temperature, amount of water, amount of solid, type of container, or exposure to light affect the growth and quality of crystals?
- What are some real-life applications or examples of crystal growth?

Suggestions For Related science projects:

- Try making crystals with other types of salts, such as alum, Epsom salt or borax. Compare their properties with table salt crystals.
- Try making crystals with different types of sugars, such as brown sugar, powdered sugar or honey. Compare their properties with white sugar crystals.
- Try making crystals with different types of liquids, such as vinegar, lemon juice or milk. Compare their properties with water crystals.
- Try making crystals with different shapes or patterns by using molds, strings or paper clips. Compare their properties with natural crystals.

Notes :

Title: "Fiber Fun: Exploring the Microscopic World of Textile Materials"

Objectives:

- To learn how to prepare and observe microscope slides of different types of fibers.
- To compare and contrast the appearance and properties of natural and synthetic fibers.
- To understand how fibers are used for various purposes and applications.

Materials and Tools:

- Microscope (preferably with a magnification range of 40x to 400x)
- Glass slides and cover slips
- Forceps or tweezers
- Water dropper
- Scissors or razor blade
- Paper towels
- Petroleum jelly (optional)
- Samples of different types of fibers, such as cotton, wool, silk, linen, hemp, jute, rayon, polyester, nylon, acrylic, etc. You can use fabric scraps, yarn, thread, or any other source of fibers.

Instructions:

1. Choose a fiber sample that you want to examine. Cut a small piece of it with scissors or a razor blade. Make sure the piece is thin enough to allow light to pass through it.
2. Place the fiber piece on a clean glass slide. Use forceps or tweezers to handle the fiber and avoid touching it with your fingers.
3. Add a drop of water on top of the fiber piece. This will help to spread out the fiber and make it easier to observe.
4. Carefully place a cover slip over the fiber piece and the water drop. Hold the cover slip at an angle and lower it slowly to avoid trapping air bubbles. Use a paper towel to blot any excess water at the edges of the cover slip. You can also seal the edges with petroleum jelly to prevent the slide from drying out.
5. Repeat steps 1 to 4 for each fiber sample that you want to examine.
6. Place one slide on the microscope stage and secure it with the clips. Adjust the light source and the focus knobs until you get a clear image of the fiber. Start with the lowest magnification objective (usually 4x) and then switch to higher magnification objectives (10x or 40x) as needed.
7. Observe and record the characteristics of the fiber, such as its shape, color, texture, structure, and any other features that you notice. You can also draw sketches or take pictures of what you see.
8. Compare your observations with those of other fibers and try to identify which ones are natural and which ones are synthetic. You can also use online resources or books to help you with fiber identification.

Safety Precautions:

- Be careful when handling sharp objects such as scissors or razor blades. Cut away from your body and use a cutting board or mat.
- Do not touch the microscope lenses or light source with your fingers. Use lens paper or a soft cloth to clean them if needed.
- Wash your hands after handling the fiber samples. Some fibers may cause irritation or allergic reactions.

Discussion Questions:

- What are some similarities and differences between natural and synthetic fibers?
- How do you think the structure and properties of fibers affect their function and performance?
- What are some advantages and disadvantages of using natural or synthetic fibers for different purposes?
- How do you think fibers are made or processed from their raw materials?
- What are some environmental impacts of using natural or synthetic fibers?

Suggestions For Related science projects:

- Test how different types of fibers react to various substances, such as acids, bases, stains, heat, etc.
- Measure how strong or elastic different types of fibers are by applying different forces or stretching them.
- Experiment with dyeing different types of fibers with natural or synthetic dyes and observe how they absorb or resist color.

Title: "Snowflake Science: Exploring the Beauty and Diversity of Snow Crystals

Objectives:

- To collect and observe snowflakes under a microscope
- To learn about the formation and structure of snowflakes
- To compare and contrast different types of snowflakes

Materials and Tools:

- A cold day with snowfall
- A chilled microscope slide (put it in the fridge for a few hours)
- A microscope
- A smartphone for taking pictures (optional)

Instructions:

1. Go outside with your chilled microscope slide and hold it horizontally where snowflakes can land on it. Try to avoid windy areas or places where the snow is disturbed by people or animals.
2. Wait for some snowflakes to accumulate on your slide. Do not touch them with your warm breath or hands, as they will melt quickly.
3. Bring the slide indoors and place it under the microscope. Adjust the focus and magnification until you get a clear image of the snowflakes. Observe the details of their structure and symmetry. You can also take a picture if you want to keep a record of them.
4. Repeat steps 1 to 3 with different slides and compare them. How are they similar or different? Can you identify any patterns or categories of snowflakes?
5. When you are done, return the slides to the fridge or freezer to preserve the snowflakes for later observation.

Safety Precautions:

- Do not eat or drink anything that has been in contact with snow, as it may contain pollutants or bacteria.

Discussion Questions:

- How do snowflakes form? What factors affect their shape and size?
- Why are snowflakes symmetrical? Why do they have six sides?
- What are some of the types of snowflakes? How can you classify them?
- How can you tell if a snowflake is fresh or old?
- How do snowflakes affect the climate and the environment?

Suggestions For Related science projects:

- Make artificial snowflakes by growing crystals from different substances, such as salt, sugar, borax, or alum. Compare them with natural snowflakes and see how they are similar or different.
- Investigate how temperature affects the shape and size of snowflakes by collecting them from different locations or at different times of the day. Record your observations and make graphs or charts to show your results.
- Experiment with different ways of preserving snowflakes, such as using hairspray, glue, wax, or resin. Evaluate how well they retain their shape and structure over time.

Notes :

Title: "Fishy Scales: Exploring the Structure and Function of Fish Scales"

Objectives:

- To observe and compare different types of fish scales under the microscope
- To learn about the structure and function of fish scales
- To understand how fish scales are related to fish adaptation and evolution

Materials and Tools:

- Compound microscope
- Microscope slides and coverslips
- Scalpel or scissors
- Forceps or tweezers
- Dropper Water or glycerin
- Paper towels or tissue
- Fish scales from different fish species (fresh or preserved)
- Magnifying glass (optional)

Instructions:

1. Choose a fish scale from one of the fish species and examine it with a magnifying glass. Note its shape, size, color, and texture. Record your observations in a table or a notebook.
2. Use a scalpel or scissors to cut a small piece of the scale. Be careful not to cut yourself or damage the scale.
3. Place the scale piece on a microscope slide and add a drop of water or glycerin on top of it. Use a dropper or pipette to avoid spilling.
4. Use forceps or tweezers to gently lower a coverslip over the scale piece. Try to avoid trapping air bubbles under the coverslip.
5. Place the slide on the microscope stage and adjust the light source and focus. Start with the lowest magnification and then increase it gradually until you can see the scale clearly.
6. Observe the scale under the microscope and look for its features, such as the outer layer (epidermis), the middle layer (dermis), and the inner layer (hypodermis). You may also see growth rings, pigment cells, blood vessels, nerves, and other structures. Sketch what you see and label the parts of the scale.
7. Repeat steps 1 to 6 with scales from other fish species. Compare and contrast the scales from different fish and note their similarities and differences.
8. Clean up your work area and dispose of the scales properly. Wash your hands thoroughly after handling the scales.

Safety Precautions:

- Wear gloves, goggles, and apron when handling fish scales to avoid contact with bacteria, parasites, or toxins that may be present on them.
- Wash your hands before and after touching fish scales or any other materials used in this activity.
- Do not eat or drink in the laboratory or near the fish scales.
- Handle sharp objects such as scalpel or scissors with care and store them safely when not in use.
- Do not touch your eyes, nose, mouth, or any other body parts with your hands while working with fish scales.

Discussion Questions:

- What are some functions of fish scales?
- How do fish scales help fish survive in their habitats?
- What are some types of fish scales and how do they differ in structure and appearance?
- How can you tell the age of a fish by looking at its scales?
- How are fish scales related to fish classification and evolution?

Suggestions For Related science projects:

- Experiment with different methods of preserving fish scales, such as drying, freezing, salting, or pickling.

Notes :

Title: "How old is that fish? A scale-reading activity."

Objectives:

- To learn how to collect and prepare fish scales for age determination.
- To observe and count the growth rings on fish scales under a microscope.
- To compare the growth patterns of different fish species and sizes.

Materials and Tools:

- Fish samples of different species and sizes (fresh or frozen).
- Forceps or tweezers.
- Envelopes or paper bags for storing scales.
- Microscope slides and cover slips.
- Microscope (preferably with a magnification of at least 40x).
- Ruler or measuring tape.
- Scale or weighing machine.
- Pencil and paper for recording data.

Instructions:

1. *Choose a fish sample and measu*re its length and weight. Record the data on a paper.
2. Locate the scale pocket on the side of the fish, behind the gill cover and above the lateral line. Use forceps or tweezers to gently remove a few scales from this area. Try not to damage the scales or the fish skin.
3. Place the scales in an envelope or paper bag and label it with the fish species, size, and date of collection. Repeat steps 1 to 3 for each fish sample.
4. Prepare a microscope slide by placing a drop of water on it. Carefully transfer one scale from the envelope or bag to the slide using forceps or tweezers. Cover the scale with a cover slip and press gently to flatten it. Avoid creating air bubbles under the cover slip.
5. Examine the scale under a microscope at a magnification of at least 40x. Adjust the focus and lighting until you can see the dark and light bands radiating from the center of the scale. These bands are called circuli and they indicate the growth rate of the fish. The wider the spacing between the circuli, the faster the growth. The narrower the spacing, the slower the growth.
6. Count the number of paired dark and light bands, or annuli, from the center to the edge of the scale. Each annulus represents one year of growth.. Record the age on a paper along with the fish species and size.
7. Repeat steps 4 to 6 for each scale sample. Compare your results with your classmates and discuss any differences or similarities.

Safety precautions:

- Wash your hands before and after handling fish samples.
- Wear gloves if you have any cuts or allergies on your hands.
- Be careful not to prick yourself with the forceps, tweezers, or fish spines.
- Dispose of any leftover fish samples or scales in a proper bin.

Discussion questions:

- How did the age of the fish correlate with its size? Did larger fish tend to be older than smaller fish?
- How did the growth patterns vary among different fish species? Which species had faster or slower growth rates?
- How did the growth patterns vary within a single fish species? What factors could affect the growth rate of a fish?
- What are some advantages and disadvantages of using scales for age determination? What are some other structures that can be used for this purpose?

Suggestions For Related Science Projects:

- Investigate how environmental factors such as temperature, food availability, or pollution affect the growth rate of fish scales.
- Compare different methods of age determination such as otoliths, vertebrae, or fin spines and evaluate their accuracy and reliability.
- Explore how aging data can be used for fisheries management and conservation purposes.

Notes :

Title: "Seeing Red – A Bloody Investigation"

Objectives:

- To observe the structure and function of red blood cells in a fish fin
- To learn about the role of red blood cells in oxygen transport

Materials and Tools:

- Medium-sized goldfish
- Two microscope slides
- Dip net
- Medicine dropper or dropper bottle with water
- Compound microscope
- Two wads of cotton (one thick and one thin)

Instructions:

1. Fill a large container with water from the aquarium where the goldfish are kept. This will be used to keep the fish moist and comfortable during the experiment.
2. Soak a thin wad of cotton in water and spread it toward one end of a microscope slide. At the other end, place a clean half-slide.
3. Using a net, gently remove a fish from the aquarium and carefully place it on a microscope slide with its head and body resting on the wet cotton. Its tail should lie on one half of the slide.
4. Place a thick, soaked wad of cotton over the body of the fish. Place another slide over the tail perpendicular to the first slide.
5. Carefully transfer the slide to the microscope stage and position it such that you can observe the fin, near its tip, through the microscope using low power. You may need to adjust the light intensity by using the diaphragm.
6. Observe the movement of red blood cells through the vessels of the fin. You should be able to see tiny round or oval cells flowing in different directions and speeds. Red blood cells are responsible for carrying oxygen from the lungs or gills to the rest of the body.
7. After finding a clear view of red blood cells, switch to medium power and observe more closely. Record your observations for red blood cells in terms of shape, size, color and number.
8. Carefully remove the slides and cotton from the fish and return it to the aquarium water container. Make sure it is alive and well before releasing it back to its original aquarium.
9. Clean up your work area and wash your hands.

Discussion Questions:

- What are some similarities and differences between fish red blood cells and human red blood cells?
- How does the shape and size of red blood cells affect their function?

Suggestions For Related Science Projects:
- Comparing the shape and size of red blood cells in different animals and relating them to their oxygen needs.

Title: "Cheeky Cells: A Microscopic Exploration"

Objectives:

- To learn how to prepare a wet mount slide
- To observe and identify cheek cells under a microscope
- To understand the basic structure and function of cheek cells

Materials and Tools:

- Microscope
- Glass slides and cover slips
- Sterile swabs
- Methylene blue stain
- Dropper or pipette
- Water

Instructions:

1. Use a sterile swab to gently scrape the inside of your cheek to collect some cells.
2. Smear the collected cells onto a clean glass slide.
3. Add a drop of methylene blue stain to the slide using a dropper or pipette.
4. Place a cover slip over the stained cells on the slide.
5. Observe the slide under the microscope, starting with the lowest magnification and gradually increasing it.

Safety Precautions:

- Handle the microscope, glass slides, and cover slips with care to avoid breakage.
- Wear gloves when handling the methylene blue stain as it can stain skin and clothing.

Discussion Questions:

1. What structures can you identify in the cheek cells?
2. How does the structure of cheek cells relate to their function?
3. How do cheek cells compare to other types of cells you have observed?

Suggestions For Related Science Projects:

- Compare cheek cells from different individuals to see if there are any observable differences.
- Observe other types of cells (e.g. onion cells, plant leaf cells) and compare their structure and function to that of cheek cells.

Title: "Fingerprint Forensics: A Microscopic Investigation"

Objectives:

- To learn about the basic principles of fingerprint analysis
- To practice collecting and analyzing fingerprints using a microscope

Materials and Tools:

- Microscope
- Microscope slides
- Fingerprint powder: Finger print powder is relatively inexpensive and easily obtained online. Presently available at Adorama – online.
- Soft brush
- Clear tape
- White paper

Instructions:

1. Touch a microscope slide with your finger to leave a fingerprint.
2. Sprinkle a little bit of powder on the microscope slide, then gently swipe off the excess with a soft brush.
3. Stick a piece of clear tape over the fingerprint firmly and carefully lift up.
4. Place the tape on a piece of white paper to look at the print closely.
5. Observe the fingerprint under the microscope and analyze its characteristics.

Safety precautions:

- Handle the microscope and other tools carefully to avoid damage or injury.
- Wear gloves when handling fingerprint powder to avoid skin irritation.

Discussion questions:

- What are some common characteristics of fingerprints?
- How can fingerprints be used in forensic investigations?
- What are some limitations of using fingerprints as evidence?

Suggestions For Related Science projects:

- Collect the fingerprints from a number of different people on slides and use those fingerprints to identify a "mystery fingerprint".
- Compare the fingerprints of different individuals to see if there are any similarities or differences.
- Investigate the effectiveness of different methods for collecting fingerprints (e.g. using different types of powders or lifting techniques).
- Research famous cases where fingerprint evidence played a crucial role in solving the crime.

Title: "Hair Ye Hair Ye : A Microscopic Investigation"

Objectives:

- To observe and compare the microscopic structure of different types of hair
- To understand the role of hair in various species
- To develop skills in using a microscope and preparing slides

Materials and Tools:

- Microscope
- Glass slides and cover slips
- Scalpel or scissors
- Tweezers
- Water
- Dropper
- Hair samples (human, dog, cat, etc.)

Instructions:

1. Collect hair samples from different sources, such as human, dog, cat, etc. Make sure to label each sample.
2. Place a drop of water on a glass slide.
3. Using tweezers, carefully place a hair sample on the water droplet.
4. Place a cover slip over the hair sample.
5. Observe the hair sample under the microscope at different magnifications.
6. Repeat steps 2-5 for each hair sample.
7. Compare and contrast the microscopic structure of the different hair samples.

Safety Precautions:

- Handle scalpels or scissors with care to avoid injury.
- Follow all safety guidelines for using a microscope.

Discussion Questions:

- How do the different hair samples compare in terms of structure and appearance?
- What role does hair play in different species?
- How does the microscopic structure of hair relate to its function?

Suggestions For Related Science Projects:

- Investigate the effects of different hair treatments (e.g. dyeing, perming) on the microscopic structure of hair.
- Compare and contrast the microscopic structure of animal fur and human hair.

Title: "Hair Today, Gone Tomorrow: Examining Hair Damage"

Under a microscope, you can see the difference between healthy and damaged hair. Healthy hair will have regular, well-closed scales and appear smooth and homogeneous because it has an intact cuticle layer. Damaged hair is stripped of the cuticle layer and appears irregular and open. The hair cuticle is the outermost part of the hair shaft. It is formed from dead cells, overlapping in layers and protects the hair shaft.

Objectives:

- To learn how to use a microscope to examine hair samples
- To understand the causes and signs of hair damage
- To compare and contrast healthy and damaged hair samples

Materials and Tools:

- Microscope
- Glass slides and cover slips
- Dropper or pipette
- Water
- Hair samples (healthy and damaged)

Instructions:

1. Collect hair samples from different individuals, making sure to include both healthy and damaged hair.
2. Place a small drop of water on a glass slide.
3. Place a hair sample on the water droplet and cover it with a cover slip.
4. Place the slide on the microscope stage and focus on the hair sample using the low power objective lens.
5. Observe the hair sample and make note of its characteristics, such as its texture, color, and any signs of damage.
6. Repeat steps 2-5 for each hair sample.

Safety Precautions:

- Handle the microscope and glass slides with care to avoid injury.
- Do not touch the microscope bulb or lens with your fingers.

Discussion Questions:

1. What are some common causes of hair damage?
2. How does damaged hair differ from healthy hair in terms of its appearance and texture?
3. Can hair damage be reversed or repaired?

Suggestions For Related Science Projects:

- Investigate the effects of different hair treatments (such as coloring or heat styling) on hair health.
- Compare the effectiveness of different hair care products in preventing or repairing hair damage.

Title: "Hair Detective: Solving the Mystery of the Unknown Hair Sample"

Objectives:

- To learn how to use a microscope to examine hair samples
- To compare and contrast hair samples from different sources
- To use critical thinking and observation skills to match an unknown hair sample to its source

Materials and Tools:

- Microscope
- Glass slides and cover slips
- Forceps
- Hair samples from various sources (e.g. human, dog, cat, rabbit)
- Unknown hair sample

Instructions:

1. Collect hair samples from various sources and label them accordingly.
2. Place a small amount of each hair sample on a glass slide and cover with a cover slip.
3. Examine each hair sample under the microscope and record your observations (e.g. color, texture, thickness).
4. Compare your observations of the unknown hair sample to those of the known samples to determine its source.

Safety Precautions:

- Handle the microscope and glass slides with care to avoid injury.
- Wash your hands after handling the hair samples.

Discussion Questions:

1. What characteristics of the hair samples did you use to determine the source of the unknown sample?
2. How might this activity be useful in a real-life forensic investigation?
3. What other types of evidence could be used in a forensic investigation?

Suggestions For Related Science Projects:

- Investigate the use of DNA analysis in forensic investigations.
- Explore the use of other types of evidence (e.g. fingerprints, fibers) in forensic investigations.

Title: "Lashing Out: A Close-Up Look at Eyelashes"

Eyelashes can differ in thickness, color, length and curliness. In addition the eyelashes of Asians have been shown to differ from that of Caucasians.

Objectives:

- To learn how to use a microscope to examine small objects
- To observe and compare the structure of eyelashes from different individuals

Materials and Tools:

- Microscope
- Glass slides and cover slips
- Tweezers
- Eyelash samples from different individuals

Instructions:

1. Collect eyelash samples from different individuals using tweezers.
2. Place an eyelash on a glass slide and cover it with a cover slip.
3. Place the slide on the microscope stage and adjust the focus to view the eyelash.
4. Observe and compare the structure of the eyelashes from different individuals.

Safety Precautions:

- Handle the glass slides and cover slips carefully to avoid breakage.
- Use caution when handling the tweezers to avoid injury.

Discussion Questions:

- What differences did you observe between the eyelashes of different individuals?
- What similarities did you observe between the eyelashes of different individuals?
- How do you think the structure of an eyelash affects its function?

Suggestions For Related Science Projects:

- Compare the structure of eyelashes from different species of animals.

Title: "Feather Frenzy: Investigating the Structure of Feathers"

The barbs are the branches that grow from the central shaft (rachis) of a feather. The barbules are mini-bards that grow from the central shaft of each barb. The barbules on one side of the shaft are smooth while those on the other side have tiny little hooks called barbicels that grab the smooth barbules that lie next to it

Objectives:

- To observe and identify the different parts of a feather
- To understand the function of each part of a feather
- To compare and contrast the structure of feathers from different bird species

Materials and Tools:

- Microscope
- Slides and cover slips
- Dropper
- Feathers from different bird species
- Tweezers
- Scissors

Instructions:

1. Collect feathers from different bird species. Make sure to handle them with care to avoid damaging them.
2. Using scissors, carefully cut a small section of the feather and place it on a microscope slide.
3. Add a drop of water to the feather section using a dropper and cover it with a cover slip.
4. Observe the feather section under the microscope. Try to identify the different parts of the feather such as the shaft, barbs, and barbules.
5. Repeat steps 2-4 with feathers from different bird species and compare their structures.

Safety Precautions:

- Handle sharp tools such as scissors and tweezers with care to avoid injury.
- Wash your hands after handling feathers to avoid any potential contamination.

Discussion Questions:

1. What are the different parts of a feather and what is their function?
2. How do the structures of feathers from different bird species compare?
3. How does the structure of a feather help a bird fly?

Suggestions For Related Science Projects:

1. Investigate how the color of feathers is produced.
2. Explore how feathers provide insulation for birds.
3. Research how birds use their feathers for communication and display.

Title: "Sleepy Bugs: Anesthetizing Insects with Common Materials"

Objectives:

- Learn how to safely anesthetize insects for observation and study
- Understand the effects of different anesthetics on insects

Materials and Tools:

- Insect-proof containers
- Forceps for transferring insects
- CO2 or a volatile liquid anesthetic (e.g. ethanol or isopropanol)
- Glass beakers/vials for test liquids
- Eye dropper
- Cotton swap for topical application

Instructions:

1. Prepare the anesthetic: If using CO2, ensure that the gas is ready for use. If using a volatile liquid anesthetic, pour a small amount into a glass beaker or vial.
2. Transfer the insects to be anesthetized into an aerated insect-proof container.
3. Anesthetize the insects: If using CO2, gently introduce the gas into the container. If using a volatile liquid anesthetic, dip one end of an eye dropper into the liquid and then carefully release a small amount of the liquid onto a piece of tissue paper or cotton ball placed inside the container.
4. Observe the insects as they become anesthetized.

Safety Precautions:

- Handle insects gently and with care to avoid harming them.
- Use caution when handling volatile liquid anesthetics and follow all safety guidelines.

Discussion Questions:

- How do different anesthetics affect insects?
- What are some potential applications for anesthetizing insects?

Suggestions For Related Science Projects:

- Investigate the effects of different concentrations of anesthetics on insects.
- Compare the effectiveness of different types of anesthetics on various insect species.

Title: "Mosquitoes Up Close: A Compound Microscope Investigation"

Objectives:

- To learn how to prepare and view mosquito specimens under a compound microscope
- To observe and identify the different parts of a mosquito's anatomy
- To understand the role of mosquitoes in the ecosystem and their impact on human health

Materials and Tools:

- Compound microscope
- Glass microscope slides and coverslips
- Mosquito specimens (dead)
- Dissection tools (scalpel, tweezers, scissors)
- Dropper bottle with water or alcohol
- Lens paper or tissue

Instructions:

1. Set up the compound microscope according to the manufacturer's instructions.
2. Using dissection tools, carefully place a dead mosquito specimen onto a glass microscope slide.
3. Add a drop of water or alcohol onto the specimen to help it adhere to the slide.
4. Carefully place a coverslip over the specimen, avoiding air bubbles.
5. Place the slide onto the microscope stage and secure it with the stage clips.
6. Using the lowest power objective lens, focus on the specimen and adjust the lighting as needed.
7. Observe the different parts of the mosquito's anatomy, such as its head, thorax, abdomen, wings, and legs.
8. Switch to higher power objective lenses to view finer details of the specimen.
9. Record observations and make sketches of what is seen under the microscope.

Safety precautions:

- Handle dissection tools with care to avoid injury.
- Use caution when handling dead mosquito specimens to avoid contamination.

Discussion questions:

- What are some distinguishing features of mosquitoes that can be observed under a compound microscope?
- How do mosquitoes feed and what is their impact on human health?
- What role do mosquitoes play in their ecosystem?

Suggestions For Related Science Projects:

- Investigate the life cycle of mosquitoes by observing different stages of development under a microscope.
- Compare and contrast mosquito anatomy with that of other insects.
- Research and present on mosquito-borne diseases and their impact on human populations.

Title: "Winging It: Investigating Insect Wings with a Compound Microscope"

Objectives:

- To learn how to use a compound microscope to observe insect wings
- To compare and contrast the wing structures of different insects
- To understand the function and importance of insect wings

Materials and Tools:

- Compound microscope
- Glass slides and cover slips
- Insect wings (from various insects)
- Forceps
- Dropper
- Water or mounting medium

Instructions:

1. Prepare the insect wings by carefully removing them from the insect body using forceps.
2. Place a drop of water or mounting medium on a glass slide.
3. Place the insect wing on the drop of liquid and carefully cover it with a cover slip.
4. Place the slide on the microscope stage and adjust the focus to observe the wing structure.
5. Repeat with wings from different insects and compare their structures.

Safety Precautions:

- Handle the microscope and glass slides with care to avoid breakage.
- Use forceps carefully to avoid injury.

Discussion Questions:

- How do the wing structures of different insects vary?
- What is the function of insect wings?
- How do insect wings help them survive in their environment?

Suggestions For Related Science Projects:

- Investigate how wing structure affects flight in different insects.
- Compare the wing structures of insects from different habitats.
- Research how insects use their wings for purposes other than flight.

Title: "Flea Frenzy: A Close-Up Look at Fleas"

Fleas are incredible creatures. Fleas can jump 30,000 times in a roe. Fleas can lift object 150 times heavier than their body weight. An adult flea can lay 50 eggs per day and can eat as many as 50 blood meals in one day.

Objectives:

- To learn how to use a compound microscope to observe small specimens
- To observe and identify the physical characteristics of fleas
- To understand the life cycle and behavior of fleas

Materials and Tools:

- Compound microscope
- Glass slides and cover slips
- Dropper or pipette
- Flea specimens (can be obtained from a pet or purchased from a biological supply company)
- Tweezers
- Methylene blue stain (optional)

Instructions:

1. Set up the compound microscope according to the manufacturer's instructions.
2. Using tweezers, carefully place a flea specimen on a glass slide.
3. If desired, add a drop of methylene blue stain to the specimen to enhance visibility.
4. Carefully place a cover slip over the specimen, taking care not to trap any air bubbles.
5. Place the slide on the microscope stage and adjust the focus and magnification to observe the flea in detail.
6. Observe and record the physical characteristics of the flea, such as its size, shape, color, and any distinguishing features.
7. Repeat with additional flea specimens if desired.

Safety Precautions:

- Always handle the microscope and glass slides with care to avoid injury.
- Wear gloves when handling flea specimens to avoid contact with potential allergens or pathogens.

Discussion Questions:

1. What physical characteristics did you observe in the flea specimens?
2. How do these characteristics help fleas survive in their environment?
3. What is the life cycle of a flea? How does it differ from other insects?
4. How do fleas impact their host organisms?

Suggestions For Related Science Projects:

- Investigate the effectiveness of different flea treatments on flea populations.

Title: "Fruit Fly Frenzy: Observing Drosophila melanogaster under the Microscope"

To catch fruit flies - Put some bait inside a glass jar. You can use ketchup, overripe vegetables, or a fermented drink like apple cider vinegar, beer, or wine.

Then, put a paper funnel over the jar's opening with the spout pointed downward to provide a small opening that is simple for flies to enter but not to leave. After capturing - Put the container in the freezer to kill the flies for observation.

Objectives:

- To learn how to use a compound microscope to observe small specimens
- To observe the anatomy and behavior of fruit flies
- To understand the importance of fruit flies in scientific research

Materials and Tools:

- Compound microscope
- Glass microscope slides and coverslips
- Fruit flies (Drosophila melanogaster)
- Small paintbrush or feather
- Dropper bottle with water

Instructions:

1. Set up the compound microscope according to the manufacturer's instructions.
2. Use a small paintbrush or feather to gently transfer a fruit fly onto a glass microscope slide.
3. Add a drop of water to the slide to help keep the fruit fly in place.
4. Carefully place a coverslip over the fruit fly, being careful not to crush it.
5. Place the slide on the microscope stage and adjust the focus and magnification to observe the fruit fly.
6. Observe the fruit fly's anatomy and behavior, taking notes and sketches as desired.

Safety Precautions:

- Always handle the microscope and glass slides with care to avoid injury.
- Be gentle when transferring the fruit fly to avoid harming it.

Discussion Questions:

- What anatomical features of the fruit fly were you able to observe?
- How did the fruit fly behave when placed on the microscope slide?
- Why are fruit flies commonly used in scientific research?

Suggestions For Related Science Projects:

- Investigate the life cycle of fruit flies by observing them at different stages of development.

Title: "Fly Investigation: A Compound Microscope Adventure"

Flies are really interesting creatures to examine. Listed below are a few items to look at. Don't forget to focus in on the fly's legs. They are all covered with hair and at the end of each leg there are two hooks.. The hooks allow the fly to hold onto surfaces that they land on. In addition look at the pad between the hooks on their feet. This pad secretes a sticky fluid that allows the fly to stick to thing like windows when they land.

Objectives:

- To observe the physical characteristics of house flies using a compound microscope.
- To learn how to prepare and view specimens using a compound microscope.
- Identify parts on the body of the house fly
- **Compound eyes:** House flies have a pair of large complex eyes that cover most of their head. Each eye is composed of 3,000 to 6,000 simple eyes.
- **Proboscis:** For tasting and consuming meals, house flies use their proboscis, a plunger-like appendage that extends from the bottom of the head.
- **Antennae:** Houseflies depend on their keen sense of smell, provided by their antennae

Materials and Tools:

- Compound microscope
- Glass slides and cover slips
- Dropper
- Tweezers
- Methylene Blue
- House fly specimen

Step-by-step Instructions:

1. Prepare the microscope by plugging it in and turning on the light source.
2. Place a drop of water on a glass slide using a dropper.
3. Using tweezers, carefully place the house fly specimen on the water droplet on the slide. (Optional)Carefully place a cover slip over the specimen, avoiding air bubbles.
4. Place the slide on the microscope stage and secure it with the stage clips.
5. Using the coarse focus knob, bring the specimen into focus.
6. Observe the specimen under low power magnification and then switch to high power magnification if possible for more detailed observation.
7. Try it again adding a little Methylene Blue to the water solution

Safety Precautions:

- Handle the microscope and glass slides with care to avoid breakage.
- Wear gloves when handling specimens and stains.

Discussion Questions:

1. What physical characteristics of house flies were you able to observe using the compound microscope?
2. How does the use of a stain enhance visibility of the specimen?

Suggestions For Related Science Projects:
- You can make an ant farm at home. Use two glass jars with lids, one large and one small that just fits inside the larger jar. Place dirt in the space between the small jar and the large jar. The ants will build tunnels and lay eggs in the soil between the two jars.
- The soil should be a mixture of soil and sand mixture for the ants to dig and tunnel in. Try to use the same soil that the ants are using where you collect them. If you're planning to source the ants from your yard or a nearby area, your best bet is to use dirt they already naturally live in. Mix 2 parts dirt and 1 part sand before putting it in the jar.
- Once you have your jars and soil mixture ready, you can find an anthill and collect some ants to add to your farm. In addition to the ants, place some food for the ants in the jar. A little bread, fruit or sugar mixed with water is usually good. Change the food in the farm on a regular basis and keep your fingers crossed.

Notes :

Title: "Ants Under the Lens: A Compound Microscope Exploration"

Objectives:

- To observe and identify the external anatomy of an ant
- To compare and contrast different species of ants
- To understand the role of ants in their ecosystem

You can observe many structures on an ant's body, each with its own special function. For example, the ant's second body segment, the **mesosoma**, is packed full of muscles that power its three pairs of legs. The **gaster** contains the ant's heart, digestive system, and chemical weaponry. The gaster is the last segment of the ant's body. Ants also have two large compound eyes and a set of simple eyes called ocelli that detect light and shadow. They also have 2 antennae that they use to find their mates and also detect enemies.

Materials and Tools:

- Compound microscope
- Glass slides and cover slips
- Forceps
- Dropper
- Water
- Ants (different species if possible)
- Paper and pencil for recording observations

Instructions:

1. Collect ants from different locations using forceps and place them in a container with a lid.
2. Prepare a wet mount slide by placing a drop of water on the center of the slide.
3. Using forceps, carefully pick up an ant and place it on the water droplet on the slide.
4. Place a cover slip over the ant, being careful not to trap any air bubbles.
5. Observe the ant under the microscope, starting with the lowest magnification and gradually increasing it.
6. Record your observations, including the ant's body parts and any distinguishing features.
7. Repeat steps 3-6 with different ants, comparing and contrasting their anatomy.

Safety Precautions:

- Handle the microscope and glass slides with care to avoid injury.
- Be careful when collecting ants to avoid being bitten or stung.

Discussion Questions:

1. What differences did you observe between different species of ants?
2. How do the body parts of an ant help it survive in its environment?
3. What role do ants play in their ecosystem?

Suggestions For Related Science Projects:

- Investigate the behavior of ants by setting up an ant farm and observing their activities.
- Study the social structure of an ant colony by observing how ants interact with each other.

Notes :

Title: "Brine Shrimp and Microscopes: A Closer Look"

Brine shrimp are crustaceans. They belong to the phylum Arthropoda. The have jointed legs and an exoskeleton (a shell on the outside of their body).They also have 11 pair of pleopods or leg-like appendages. If you look at their legs you can see that they are designed for swimming. In addition their eyes are compound eyes – eyes that produce multiple images that are transmitted to the brine shrimp's brain. Compound eyes make movement very detectible and make the brine shrimp well aware of its environment and enemies

Brine shrimp can be raised very easily at home or in the classroom with a minimal amount of equipment. Brine shrimp eggs can be purchased at most pet shops and directions to hatch and raise them will be included. Brine shrimp eggs will hatch eggs within 24-36 hours.

Objectives:

- To learn how to use a compound microscope
- To observe the anatomy and behavior of brine shrimp
- To understand the importance of brine shrimp in aquatic ecosystems

Materials and Tools:

- Compound microscope
- Glass slides and cover slips
- Dropper or pipette
- Brine shrimp sample
- Well slide

Instructions:

1. Set up the compound microscope according to the manufacturer's instructions.
2. Place a small amount of brine shrimp in a petri dish.
3. Use a dropper or pipette to transfer a few brine shrimp onto a glass well slide.
4. Place a cover slip over the brine shrimp on the slide.
5. Place the slide on the microscope stage and adjust the focus to observe the brine shrimp.
6. Observe the anatomy and behavior of the brine shrimp and record your observations.

Safety Precautions:

- Handle the microscope and glass slides with care to avoid injury.
- Follow all safety instructions provided by the manufacturer of the microscope.

Discussion Questions:

- What did you observe about the anatomy and behavior of brine shrimp?
- How do brine shrimp contribute to aquatic ecosystems?
- What other organisms might you observe using a compound microscope?

Suggestions For Related Science Projects:

- Investigate the life cycle of brine shrimp
- Compare the anatomy of different aquatic organisms using a compound microscope
- Explore the effects of environmental changes on brine shrimp populations

Notes :

Title: "Daphnia Discovery: A Microscopic Adventure"

Daphnia are interesting creatures to watch. Even under low power you can watch the daphnia feed, watch their eye movement, and see their heart beating. Their heart is located along the back of the animal.

Daphnia are readily available online on websites like Ebay and Amazon. In addition pet shops very often sell live daphnia as a fish food. Last but not least, if you are very adventurous, you can collect and separate daphnia from a sample of collected pond water.

Daphnia can be raised in a tank or container with a large surface area and not too much depth. The water should be chlorine-free and aged. They can be fed green algae as well as a pinch or two of active yeast a day. 10%-20% of the water should be changed every week and the added water should be chlorine free. It takes about a week for undisturbed water to release almost all of its chlorine.

Objectives:

- To observe the anatomy and behavior of daphnia using a compound microscope.
- To learn how to prepare a wet mount slide for microscopic observation.
- To practice using a compound microscope to view specimens.

Materials and Tools:

- Compound microscope
- Glass slides and cover slips
- Dropper or pipette
- Daphnia culture
- Pond water or dechlorinated tap water

Instructions:

1. Fill a small container with pond water or dechlorinated tap water.
2. Use a dropper or pipette to transfer a few daphnia from the culture into the container.
3. Place a drop of water from the container onto a glass slide.
4. Use the dropper or pipette to transfer one daphnia onto the slide, in the drop of water.
5. Carefully place a cover slip over the drop of water, being careful not to trap any air bubbles.
6. Place the slide onto the stage of the compound microscope and secure it with the stage clips.
7. Use the low power objective lens to locate the daphnia on the slide.
8. Once you have located the daphnia, switch to a higher power objective lens to observe its anatomy and behavior in greater detail.

Safety Precautions:

- Handle glass slides and cover slips carefully to avoid breakage.
- Be careful when using sharp tools such as scalpels or tweezers.

Discussion Questions:

1. What anatomical features of daphnia were you able to observe?
2. How did the daphnia behave when placed on the slide?
3. How does the anatomy of daphnia relate to its behavior and habitat?

Suggestions For Related Science Projects:

- Investigate how different environmental factors (such as temperature or light) affect the behavior of daphnia.
- Compare and contrast the anatomy and behavior of different species of daphnia.

Notes :

Title: "Daphnia's Heartbeat: The Effect of Water Temperature"

Daphnia are readily available online on websites like Ebay and Amazon. In addition pet shops very often sell live daphnia as a fish food. Last but not least, if you are very adventurous, you can collect and separate daphnia from a sample of collected pond water.

Objectives:

- To observe the effect of water temperature on the heartbeat of a daphnia using a compound microscope.
- To understand the basic anatomy and physiology of a daphnia.
- To learn how to properly use a compound microscope.

Materials and Tools:

- Compound microscope
- Glass slide
- Cover slip
- Dropper
- Daphnia specimen
- Pond water
- Thermometer
- Beaker
- Ice chips

Instructions:

1. Obtain a water sample of room temperature water containing daphnia
2. Use a thermometer to measure the temperature of the water.
3. Place a drop of the water on a glass slide.
4. Place a cover slip over the daphnia specimen.
5. Place the slide onto the stage of the compound microscope and adjust the focus until the daphnia is clearly visible.
6. Observe and record the heartbeat of the daphnia.
7. Drop a small piece of ice in the original daphnia water sample and record the temperature of the water after the ice has melted
8. Repeat the above procedure several times and record the heart rate of the daphnia in each of the cooler water samples.
9.

Safety Precautions:

- Follow proper laboratory safety protocols.

Discussion Questions:

- What did you observe about the effect of water temperature on the heart beat of the daphnia?
- How does this compare to what you know about other organisms?
- What other factors might affect the heart rate of a daphnia?

Suggestions For Related Science Projects:

- Investigate how other environmental factors (such as pH or light) affect the heart rate of a daphnia.
- Compare the effect of water temperature on the heart rates of different species of daphnia or other small aquatic organisms.

Notes :

Title: "Observing Amoeba: A Microscopic Adventure"

An amoeba is a type of unicellular organism with the ability to alter its shape, primarily by extending and retracting pseudopods. When observing the amoeba you will be able to observe a thin skin that surrounds the Amoeba's body called a cell membrane. In addition you should be able to see the nucleus of the cell and also some food vacuoles. The food vacuoles look like bubbles inside the cell that contain things that the amoeba has eaten. You can also watch the amoeba move. It moves like a blob projecting portions of its body out and in the direction that it wants to move. It also surrounds whatever it is eating with these projections and then encloses what it has eaten in a newly formed food vacuole. The projection or "false feet" are called pseudopods.

The best way to collect amoeba is to gather some pond water and pond weed and allow it to it to decay in a shady warm place . When the pond weed decays and settles to the bottom or rises to the top collect a small portion and look at it under the microscope. It should contain some ameobas.

Objectives:

- To observe the structure and behavior of an amoeba under a compound microscope.
- To identify and describe the parts of an amoeba and their functions.

Materials and Tools:

- Compound microscope
- Glass slides and coverslips
- Dropper
- Pond water sample containing amoeba
- Methylene blue stain (optional)

Instructions:

1. Set up the compound microscope on a flat surface and turn on the light source.
2. Using a dropper, place a drop of pond water containing amoeba onto a glass slide.
3. (Optional) Add a drop of methylene blue stain to the water sample to make the amoeba easier to see.
4. Carefully place a coverslip over the water sample on the slide.
5. Place the slide onto the microscope stage and secure it with the stage clips.
6. Using the low power objective, focus on the water sample until you can see the amoeba.
7. Observe the amoeba's behavior and structure, taking note of its shape, movement, and any visible parts.

Safety Precautions:

- Handle the microscope and glass slides with care to avoid damage or injury.
- Use caution when handling stains or other chemicals.

Discussion Questions:

1. What parts of the amoeba were you able to see under the microscope?
2. How does an amoeba move?
3. How does an amoeba obtain food?
4. What is the function of each part of an amoeba?

Suggestions For Related Science Projects:

- Observe other microscopic organisms found in pond water, such as paramecium or euglena.
- Compare and contrast the structure and behavior of different types of single-celled organisms.

Notes :

Made in the USA
Coppell, TX
20 December 2023